the $39 MUSTACHE COMB:

THE START-UP GUIDE TO MANUFACTURING

BAILEY BRISCOE JONES

Published by Forge Arbor Books.

Cover & book design by Gino Verna.

ISBN: 978-0-9992068-9-8

CONTENTS

Preface

The physical products that we see online and on store shelves have navigated a tortuous path to get there. I've been involved in the design, engineering, and manufacturing of such products for most of my career. As the founder of Bright Product Development, I see many proposed consumer products in their early stages of development. I have seen products succeed and generate substantial profits, and I have seen products fail. Some fail because they do not address a need in the market. This grave error is the result of inadequate market research to validate that need. Market validation is the first hurdle—and one that we will leave for another book to explore in detail. Conversely, some products fail as a result of improper planning for the time, budget, and engineering that manufacturing requires. This book focuses on the nuts-and-bolts of product development including scheduling, budgeting, engineering, prototyping, and production. You will find specific guidance here for tackling these elements.

In addition to basic manufacturing and planning information, the book contains client and colleague interviews, case studies, and manufacturing tips from real-life projects. I hope to pass along some of the advantages of hindsight with these examples. I have also collected useful business resources at the end of the book, and you can find additional information and expanded material at brightpd.com. My purpose is to smooth the road to manufacturing, which is filled with bumps, turns, and detours. By anticipating these obstacles you can successfully manufacture profitable products.

1

Manufacturing: What Does It Take?

Congratulations! You've crafted the finest (and only!) Internet-connected mustache comb. And you've managed to put together an awesome prototype that works every time. Then you built three more just to prove you could do it. If you had to, you could probably build twenty of them. But, could you build a thousand? How about one hundred thousand?

These different volume categories each have their own challenges. It takes an inventive ingenuity to conceive an idea, as well as some craftiness to build a few homemade prototypes. You may even go through several revisions before you are convinced your idea can work. I once engineered a high-capacity tray to catch photos coming off a self-service photo-printing kiosk. The prototype trays started out as taped-up cardboard. I went through dozens of these hand-built prototypes before the tray prevented all the high-speed prints from dumping out on the floor.

Now, building a thousand of something has a different set of challenges. For example, once I was convinced that the tray could catch all the prints, I redesigned the hand-crafted prototype to reflect established sheet metal manufacturing processes. This meant that I used a readily available sheet metal thickness and designed the bends so that they could be easily formed on standard shop equipment. Any local shop could then produce hundreds or thousands of these parts according to the new design. Crowdfunded projects often provide a window that reveals the difficulties of producing quantities in the thousands. Schedules slide, production funding gets stretched, and some projects fold—all in spite of functioning prototypes and overflowing optimism. It can be a giant engineering leap from a prototype to a design for standard manufacturing processes. And, furthermore, production in the thousands is often not high enough to reach an efficiency of scale that leads to reasonable consumer prices.

Products that are meant to be manufactured at high volumes should be designed with these standard manufacturing processes in mind from the beginning. And those processes require a heavy upfront investment in special manufacturing hardware, called tooling, that aids in speedy and effective production. Even with careful design, you can expect adjustments as you ramp up to full-scale production. These adjustments come after running hundreds of test parts. Remember that expensive factory tooling you paid for? You'll likely find that you need to go back and adjust that too. This takes time and money, and it is part of the normal development process.

What does it take to get that first lot of two thousand on the shelves? "You sort of have to make two thousand of something to find out if you can make two thousand of something," says Jack Daniels, president of Eastbridge Engineering, which specializes in quoting, sourcing, and overseeing the production of consumer goods in the Asia-Pacific.

Daniels goes on to say,

> The learning curve associated with building a new product is painful and costly. While you (the entrepreneur, garage genius, maker, hacker, etc.) can handcraft one to five units on a limited budget, a contract manufacturer cannot. Even after the tooling is made and paid for (and yes, once we get past the modeling and prototype phase, tooling is a must) the contract manufacturer must use conventional and real industrial manufacturing techniques to make retail-ready products. This is an essential stage in the new product development cycle.
>
> There are very few products that are a "drop-in" fit in most factories. Until the fabricator develops the methods to make your product and then adjusts their processes to suit it, it's a learning process. . . Sometimes hundreds, and in some cases thousands [of parts], are consumed in this process. . . Get your emotions and checkbook ready to be exercised.

These days you can buy a cordless drill for twenty-five dollars. How is that even possible? To reach the low price that consumers have come to expect, you must generate a large quantity of your product in a cost-effective way so that the individual price becomes low. To set up efficient production you must design your product with care, select the right materials, choose appropriate fabrication processes, and invest in expensive tooling. Is this easy? No, but it *is* possible.

So after this sober introduction, let's slow down and methodically prepare to navigate the manufacturing jungle and harness the advantages of high-volume production. In the next chapter we will break down the process into manageable steps and get more specific from there.

2

Product Development Steps and Timeline

You should plan on constant, dedicated effort for twelve to eighteen months—in some cases even longer—with a complete team, in order to go from a well-defined product idea to a finished product that is ready to ship. It is impossible to anticipate all delays and complications, so be prepared for these setbacks from the start and schedule a realistic timeline with this in mind.

Rich LeGrand, president of Charmed Labs, which develops and sells innovative robotics technology, might advocate budgeting even more time. Their Pixie robotic camera was the star of a 250,000 dollar Kickstarter campaign in 2013. Rich reflects on how long it actually takes to develop a product: "I hate to think of this, but it typically takes [us] two to three years. At the beginning, it almost always seems like it will take half the time it actually takes."

To understand what sort of roadblocks you might encounter, take a look at some of the funded design projects on Kickstarter. Scroll through the updates and check out the explanations for delays in the schedule. These are all normal, and your project will be no different. Before crowdfunding, these delays were opaque to the consumer because the process was hidden inside corporations until they were finally ready to launch. And on top of the public scrutiny that crowdfunding is subjected to, product development delays in that arena are often compounded by an inexperienced founding team and underestimated budget requirements.

Sources for delays include potential manufacturers overpromising what they can deliver; it is easy to simply underestimate the complexity of various parts that will end up requiring revisions and adjustments. And these adjustments might come after the time that you had planned for them to be in full-scale production. I backed the Tiko 3D printer on Kickstarter in March of 2015 and it was scheduled for delivery in November of that same year. At that time, the founding team was small and competent, yet inexperienced, and they had a working prototype. A critical component of their design was a plastic extruded chassis that contains precise rails for a gantry system, and their manufacturer had indicated that it should be no problem to make. They related their manufacturing difficulties in an update as that year came to a close. They had started extruding the chassis in May but it had come out as a crooked, ugly, and inaccurate mess. Finally, by the end of that year, they had achieved the chassis quality that the design required, but they were still nowhere near delivering their product as scheduled. And that was only one of many manufacturing challenges for the project. They finally delivered some partially functioning printers more than a year later (I never did receive mine), but then the company was forced to lay off their team and wind down operations in 2017 after their funding—just under three million dollars—had dwindled to almost nothing. Two years would have been a reasonable time to deliver such a product, but their planning and budgeting was so far off that they fell short. They later acknowledged that they were unprepared for how hard it would be to go from a prototype to full manufacturing, and by the time they realized they were in trouble they were unable to turn the company around. The outcome might have been different if they had set a more reasonable timeline from the beginning to allow for all the engineering and production iterations it takes to deliver a mass-produced product.

So, taking caution, let us now look more specifically at the steps and timeline for delivering such a product. This outline would be for a handheld-sized consumer product. Let's use the mustache comb as an example. First, why an Internet-connected mustache comb? Absurd! Indeed it is. (That's right; we are leaving the market analysis to another book.) Let's claim that the comb monitors mustache health and uploads the data to a personal facial hair grooming app. This indispensable tool is comprised of several parts. The body of the comb is injection-molded plastic. The design will depend on a research phase to be sure that the handle is comfortable to hold and that the teeth are spaced just right. We don't want to take anything for granted or to simply copy what has been done before. There will be a hollow in the comb that contains a small printed circuit board (PCB) and a wireless module along with a battery. We will specify a small, energy-dense battery that will fit into our limited space. We will have a waterproof cover that caps the PCB to protect it from any moisture during our morning ablutions.

We will also have to figure out how to waterproof the single tiny button and the blinking LED that are part of the design. We plan out our schedule according to this outline:

Design and Engineering	3-6 months
Validation and Testing	1-2 months
Find and Vet Manufacturers	2-3 months
Manufacture the Tooling	6-12 weeks
Produce First Parts, Adjust Design	2 weeks
Manufacture Production Parts	2 weeks
Regulatory Testing	1-2 months
Assembly and Packaging	2 weeks
Shipping and Customs	1-2 months
Fulfillment to the Customer	1 week

No doubt you are making a mental note of your own timeline. Resist the urge to squeeze down your delivery forecast to less than twelve months. Reality will surely intervene and push that release date back out.

John Kestner, the principal of Supermechanical, a company that makes connected home products, gives these pointers for managing the product development cycle:

Even if you're an organized person, there are way too many small things demanding your attention in something this complicated. You have to use a project management app with Gantt charting (Microsoft Project, OmniPlan, TeamGantt) to turn a huge project into a series of tasks, assign them to your team, and tell you where the trouble spots are. Check weekly that everyone's on track and be honest with yourself about your progress as you update your schedule. And things will always take longer than you want, but especially when you make changes mid-stream. Reduce that likelihood by minimizing the number of challenging features in a given product. Save the rest for version two.

Kestner describes his design process with an eye toward manufacturing:

> Every product, I learn something I could've done at the beginning of the process that would've saved me time, money, and grief at the end. Manufacturing is one of the constraints that you want to solve for in order to make the most elegant and complete product . . . Design for assembly, fulfillment, and marketing are also in my head when I sketch . . . It's multi-dimensional chess—fun and challenging.

Despite all the complex tasks you will have to attend to throughout the process, much of the heavy lifting to make production go smoothly starts early with design and engineering. Beginning with savvy industrial design will result in a product that is both useful to the consumer and reasonable to manufacture. This earliest phase is not necessarily concerned with individual part design, but should anticipate materials and processes. These designs are captured with three-dimensional, computer-aided design (3D CAD), realistic-looking rendered images, and physical appearance models. Usually, it will take several weeks or more to narrow the best concepts down to one cohesive design direction. There will be a push and pull among the many priorities of a product: features, appearance, marketability, complexity, cost, etc. These competing priorities should converge through a series of design concept adjustments.

Good industrial design quickly begins to merge with engineering as the ideas start to become more real. Now, the general design concepts become realized as assemblies of individually designed pieces. Mechanisms and moving parts start to take a true mechanical form, and therefore determining proper materials and processes is critical at this point. Design for manufacturing (DFM) using principles such as those detailed in this book is paramount. There is no way to sidestep the effort it takes to get to realistic prototypes or useable CAD. Useful CAD and prototypes come after painstaking engineering.

If electrical systems are a product component, those features and their behavior are now also engineered as directed by the design concepts. Industrial designers, electrical engineers, and mechanical engineers should work closely together as the aspects of the design come together. The board layout, for instance, will be dependent on both the electrical requirements and the mechanical requirements such as board size, button location, and LED placement. For instance, in the case of our mustache comb, the industrial designer will have a specific idea of where the button should go so that it will be easy to use. The mechanical engineer may find it difficult to put it there because that area of the comb is too small for it to fit. But wait, the electrical engineer discovers a specialty switch out of Japan that might work. Oh, it is expensive. Furthermore, now the entire product would be dependent on this single component. What if it goes out of production next year? These are the kinds of issues that come up during the design and engineering, and often there are no clear answers.

A variety of prototypes will be an integral part of the design and engineering process. As part of the design validation you should get working prototypes into the hands of potential customers and listen to what they have to say about it. Make adjustments to be sure that your product addresses the needs of the market. These iterations are critical in delivering successful and relevant products.

Engineering design documentation includes 3D CAD, two-dimensional (2D) drawings, electrical design Gerber files, and other files that fully define all the individual parts of the product, along with a bill of materials (BOM). These files provide the necessary information to get realistic quotes from potential manufacturers.

Locating and vetting manufactures can be a full-time job by itself. Sure, internet searches can be a starting point, but your best quality match may have an unimpressive internet presence. It is best to have a manufacturer that is a specialist in similar products. If you are making an inexpensive plastic toy, you should search for manufacturers that make inexpensive plastic toys; an injection molder that is accustomed to making medical devices may not have the most efficient processes, or the right price for your application. Be ready to pay enough to partner with a manufacturer that you trust to deliver quality products. Cost should not be the main determining factor. That initially-appealing low quote can turn into frustration and more money out of pocket in the end. Finding a manufacturer for our mustache comb may prove to be a challenge. Ideally, we would want to work with a factory that makes combs. There could be some specific processes they have learned in order to get the little teeth to come out just right. However, a comb factory may have no experience packaging and waterproofing electronic components. Even simple products have an array of requirements that can quickly narrow your search, though the perfect match rarely exists.

I recommend visiting the manufacturer. Whether close to home or overseas, the cost is almost always justified. Even today, the manufacturing relationship is just that—a relationship. You must put in effort to be treated well. Let's face it; your product may not be the greatest source of revenue for your contract manufacturer (CM), so you will have to make yourself heard in other ways. You should nurture the process with your supplier from the very beginning, and from that point on—through the quoting, first article inspection, pilot production, and delivery. It helps to have an agent on-site to advocate for your interests if you are not able to be there personally. Independent production companies can provide this service that looks after quality control and timely delivery.

It takes a long time to manufacture tooling. The tooling fabrication may be done in-house at your contract manufacturer or, more likely, the manufacturer will sub-contract it out. Why does it take so long? One, to precisely manufacture tooling from large blocks of solid steel takes several time-consuming steps. First the tooling must be designed. The factory will work from the documentation of your product as a starting point. For injection molded parts they will consider gates, runners, vents, cooling systems, and mechanisms such as lifters and slides. Then they can manufacture the tooling using various machining and finishing steps. But secondly, tooling manufacturing takes a long time because your project is likely not their number one priority and you must get in line behind previous commitments. Depending on your relationship and your good fortune, it could take from a few weeks to three months to manufacture the tooling, which often represents one of the largest capital expenses of product development. It is common practice to make partial payment up front with the remainder due upon delivery.

You or your agent should be on-site for first article inspection. When the parts first come off the line you will likely need to make small adjustments to the parts for better quality, fit, and appearance. The manufacturer will also be adjusting their production parameters for better part quality. You put effort into these first (and scrapped) parts for the sake of quality and efficient production later. You and your manufacturer should be in agreement of the acceptance criteria for the parts. This can be communicated with drawings, "golden samples," and other quality documentation. Golden samples are specially designated reference parts that are used to evaluate parts off the manufacturing line.

Let's follow the mustache comb through first article inspection. We are at the factory and they have injection molded a hundred or so of each part. They look good. But when we go to put them together, there is a small gap between the cover and the comb which renders the waterproofing seal ineffective. Fortunately, the engineer had anticipated this possibility and had designed a rib that could be adjusted in a safe and predictable way. The factory engineers take the tooling and grind off .015 of an inch (.38mm) so that the rib gets just a bit thicker. (Removing material from the tool, or mold, means that material will be added to the part.) This change had been anticipated in the design as tool-safe, also called steel-safe. That is, it anticipated removal of material from the tool. It is more difficult and costly to try to add material to a tool. So, the crew runs a new batch of parts and now all the pieces fit together as planned. Well done, team.

Many products require regulatory testing for FCC (Federal Communications Commission), UL (Underwriters Laboratories, USA), and CE (Conformité Européenne) compliance. You should start this process before your product is complete and submit it to be tested early, likely at the point of the engineering build (EB), or manufacturing build (MB), which consists of the first fully-assembled, as-manufactured examples of your product. The MB units are used for product evaluation and testing. Testing can consume several months, and UL testing alone can run well over 10,000 dollars while FCC testing can come to 15,000 dollars. Regulatory testing for a small home appliance I worked on tallied up to about 25,000 dollars.

Finally, you are ready for full-scale production, assembly, packaging, and shipping. Your contract manufacturer should be able to facilitate and integrate these steps, with you also playing a supervisory role, especially for the shipping. Stick with it. You are getting closer, but here deep into the development, it starts to take increasing effort to get seemingly smaller results.

For example, say we have specified custom colors for the mustache comb: Near Black, Tortoise, and Metallic Bronze. We did the pilot production run in an unspecified generic black since that stock color is readily available in most varieties of plastic. We determined that it would not be worth it to custom-color the slightly lighter Near Black that we had wanted, so we went with a generic black instead. However, thirty-five percent of our pre-orders were for Metallic Bronze and we are not willing to lose those sales. The problem we discovered is that this particular color additive weakens the material and the teeth of the comb tend to snap off. We have several options: go with a non-metallic bronze, paint the comb after injection molding, or change the plastic material. None of these options are guaranteed. We eliminate the first option because it is not close enough to what the customers ordered, and we eliminate the second option for fear that the paint will scratch off; we are left with the third option— change materials. It turns out that the new material has a different shrink rate than the previous material, so to maintain the same size and tolerances we would have to cut new tools. This would represent an unacceptable increase in development cost and time. So, in spite of any tolerance problems it might create, we decide to shoot the new material in the existing molds. What is our risk here? Remember our effort to adjust the mold so that we would have a watertight seal for the cover? We might have to do that all over again. And we may find that the adjustment would not be tool-safe this time around since this is a change no one anticipated. These little unexpected problems can have a big influence on the time and cost to deliver your product. Take courage that you are in good company when you encounter them.

Lane Musgrave, co-founder of Reserve Strap, a battery backup watchband, reflects on this phenomenon:

> The last mile in a marathon is the most difficult. You're trying ten times as hard and achieving one-tenth the result. That's what the final ten percent of product development feels like. The only phase of Reserve Strap development when I wanted to quit was when the product was almost complete. It seems contradictory, but when you're

in the thick of it, you don't know if/when you'll be 'done.' As soon as you solve one problem, it creates another and the endless game of whack-a-mole is the most frustrating part of design and development.

This is a time for quick and decisive decisions to resolve the issues that you can, and to dismiss the ones that you cannot. It is important to release a product that meets expectations, but many of those perfecting adjustments may be left for subsequent generations of the product.

In quickly changing industries and for products that are produced in lower quantities it can often make sense to manufacture close to home. In these instances, the shipping time from your vendor can be measured in days. For higher volume production or specialized production it is often most effective to manufacture in a worldwide manufacturing center. The shipping process from these overseas manufacturers can easily take two months altogether. Product preparation, packing, and staging at the port can take more than a week. The shipping time on the water can take about four weeks from China, for example, and it can take a week, or even much longer, to clear domestic customs. Fulfillment companies can aid in this process, and continue on from customs to their warehouses, and then to the individual customers. Or if you have your own distribution system, you would transport the product from the port to your own warehouses, and finally to the end customer.

These steps have added up to something that is starting to seem overwhelming. We'll break them down into more manageable pieces in the coming chapters, but for now, let's look to a few practical resources: Need a product design firm to get you started? Check out the design directory from Core77. Trouble with fulfillment? Outsource it to a company like Blackbox. They shipped the Exploding Kittens game of Kickstarter fame to over 200,000 backers. And there are many other choices like them. Don't know where to start with selecting a manufacturer? Use a service like Eastbridge Engineering. They will quote and vet several well-matched choices. Already drowning with the engineering of your project? Check out hardware product accelerators like the Flex Invention Lab. See the RESOURCES section at the end of this book for more information on these and many other services. You can also check out brightpd.com/resources for an updated and expanded resource list.

For a sample project schedule that includes the principles of this chapter, look toward the end of the book to APPENDIX I. This project schedule follows a typical development process with parallel paths for independent items and long lead-time items in order to compress the time to delivery. As evident from its interconnected nature, late completion of any individual step can lead to late fulfillment of the product.

In the next chapter we will take a closer look at the first item on the timeline: design and engineering. Inevitably, the design of a product will need to be suited for particular manufacturing processes. And the most suitable processes are closely tied to production volume. So, we introduce design for manufacturing (DFM): a balancing act between market requirements, features, and appearance, all countered with the cold reality of established fabrication processes and minimum production volume.

3

Design for Manufacturing: Anticipate Production Volume and Understand Tooling

The design of something that is produced in ones or tens is entirely different from that of an item produced in high quantities. To take advantage of lower individual part costs associated with higher quantities so that we can produce that twenty-five dollar drill (or our thirty-nine dollar mustache comb!), we must have efficient processes. And those processes affect the very design of the product. To put it another way, a design that is easy to 3D-print will not necessarily be possible to make with the final material such as plastic, sheet metal, wood, or anything else using real manufacturing processes. Plastic parts that are produced in large quantities must accommodate uniform wall thickness, draft angle, and many other special considerations. Likewise, formed sheet metal has its own specific design requirements. In this way, the material has a direct relationship to the design.

The production schedule of a new product typically passes through several stages during its development. The goal will be to get to the highest available volume category as quickly and efficiently as possible. The journey should include pre-production runs to verify and optimize the process before jumping into full production. Getting there and maintaining steady production contributes to lower-cost parts that allow your business to be profitable. Profitable products most often would have a minimum production volume in the high thousands per year. Most products we see and buy in stores are produced in the tens-of-thousands per year and higher. Here's a rough breakdown of production methods for different yearly volume categories:

1-10: You, your colleagues, and friends craft small proof-of-concept batches for product development, product testing, and validation.

10-100: Use any variety of prototyping methods to produce parts. Hire temporary workers for assembly and packaging. Solicit workers through Internet job boards and pay by the hour or by the part.

100-1000: Good luck. Few products could be profitable at this volume. Exceptions to the rule could include expensive custom built items like furniture or expensive medical and surgical products.

1000-5000: Use a small-scale contract manufacturer close to home or abroad. Some prototyping facilities fulfill this quantity niche with streamlined, low-cost tooling. This is a difficult volume category to fulfill—too big to do it yourself, and too small to take great advantage of the efficiency of scale. Many new companies start with products in this category. However, long-term business health usually requires growing into higher volume production and sales.

5000-50,000: Use a small to mid-scale contract manufacturer close to home or abroad. Use more efficient tooling (read: more expensive) to begin to take advantage of scale. Contract manufacturers in this volume category expect their customers to generate hundreds-of-thousands US dollars revenue per year for the factory.

50,000-250,000: Use mid-scale manufacturers in worldwide manufacturing centers. Leverage production tweaks to make the manufacturing process more efficient. Use full hardened steel tooling, multi-cavity molds and other tooling to bring down individual piece price. Contract manufacturers in this volume category expect their customers to generate millions of US dollars revenue per year for the factory.

250,000-1 million+: Use large-scale manufacturers in worldwide manufacturing centers. Request detailed quotes with individual part breakdown and negotiate terms and price for individual part procurement and manufacture. At these volumes, it pays to work for continual price reduction and to increase production efficiency. Take the initiative to negotiate lower prices as the production becomes more efficient, while also allowing your production partners to make a fair profit. Companies producing at these volumes will have departments dedicated to sustaining engineering, procurement, factory optimization, and contract negotiation. Manufacturers in this volume category expect their customers to generate tens-of-millions or more of US dollars revenue per year for the factory. These contract manufacturers include such recognizable names as Foxconn and Flextronics.

Manufacturing Methods and Materials

As we have discussed, many manufacturing processes require specialized hardware that is specific to the part being made. This is called the tooling. The tooling is what allows multiple parts to be made quickly and efficiently. The expense of the tooling is usually covered by an up-front fee to be paid to the manufacturer before production begins. The tooling cost could range from less than one hundred dollars for a simple sheet metal part to over 20,000 dollars for a complicated hand-held size plastic part. Injection molding tooling for larger plastic parts can easily run over 100,000 dollars. However, with efficient processes and tooling the individual piece price for these same parts could be as low as a few dollars or even a few cents. So the natural consequence is that the tooling price has great influence over the practical production volumes.

Figure 3.1 | Multi-cavity injection molding tooling[1]

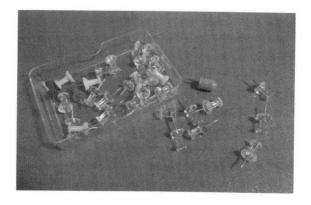

Figure 3.2 | Even cheap plastic parts require expensive tooling.

Our mustache comb has two main pieces in addition to the printed circuit board, PCB: the body of the comb and a cover to the electronics compartment. Both of these parts are injection molded. The body of the comb is the larger, but simpler part. The cover has an added complication of being injection molded in two stages. The bulk of the cap is injection molded with hard plastic and in the second stage it is overmolded with a soft, rubber-like seal. The tooling for the body of the comb comes to $5500 and the part price is $2.25. The tooling for the overmolded cover is $9750 and the part price is $1.95.

To get a better idea of the costs associated with various manufacturing processes, let's take a look at some specific examples. We'll use the study of these examples to get an idea of materials, processes, tooling costs, and piece price. I've chosen these parts as good examples of their categories; there will always be exceptions that vary widely from what is represented here.

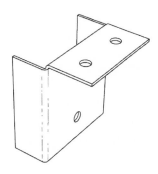

Figure 3.3

Item:	Bracket
Process:	Formed Sheet Metal
Material:	.048" (1.2mm) thick cold rolled steel
Size:	2" (50mm)
Finish:	Powdercoat
Tooling:	None
Quantity:	4000
Piece Price:	$4 USD
Country:	USA

These brackets secure the legs to a small table. Six brackets were used on each piece, which boosted the total bracket quantity to a more attractive volume for the manufacturer despite a relatively low quantity of furniture sales. (It is a recurring theme to find manufacturers decline projects on account of insufficient volume.) The pattern for the part was cut with a laser cutter, which requires no tooling. In very high volumes, this part might be stamped with a die instead to achieve faster and lower cost production. The straight-line bends with standard bend radii of this part allowed it to be formed on a sheet metal break, which is standard shop equipment and typically does not require tooling. However, each bend had to be separately inserted into the machine, which added to the cycle time, and correspondingly, the cost. A four-slide press with custom tooling would accomplish all the bends at once and thus be a more efficient method to use. At these lower volumes, however, it made sense for us go with a local sheet metal vendor and to use processes that did not require tooling.

Figure 3.4

Item:	End Table Support
Process:	Formed Sheet Metal
Material:	.25" (6.4mm) thick aluminum
Size:	12" (305mm) wide
Finish:	Powdercoat
Tooling:	None
Quantity:	1200
Piece Price:	$5.75 USD
Country:	USA

At one-quarter of an inch (6.4mm), this formed sheet metal part is at the upper bounds of what can be processed on standard shop equipment. Again, at this low quantity, the parts were cut with a laser and formed on a sheet metal break. Automated production of this part would require very heavy tooling and large, specialized equipment.

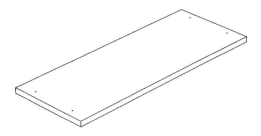

Figure 3.5

Item:	Tabletop
Process:	Milling
Material:	½" (12mm) thick Baltic Birch Plywood
Size:	30" (760mm) long
Finish:	Clearcoat
Tooling:	None
Quantity:	300
Piece Price:	$12 USD
Country:	USA

This tabletop was produced locally at a cabinet shop with computer-controlled equipment. Several processes could have been used to create the simple shape, but it was milled on a CNC milling table. The mill cut the board to shape and drilled the pilot holes to the specified depth. The automated equipment did not require tooling.

Figure 3.6

Item:	Glasses Frame
Process:	Injection Molding
Material:	Polypropylene plastic
Size:	5.7" (145mm) wide
Finish:	None
Tooling:	$11,860 USD
Quantity:	15,000
Piece Price:	$0.26 USD
Country:	Taiwan, PRC

This part would be produced using full-hardened steel tooling as the mold. In prototype quantities, a cheaper aluminum tooling (soft tooling) could be used, but that would not be appropriate at this production volume. This project was quoted with plans for a single cavity mold. That is, with each injection molding cycle, one part would be produced. A multi-cavity mold costs more but would produce several parts with each molding cycle, and have a correspondingly lower part price, since price has a direct relationship with cycle time. The most established centers for injection molding are in Asia, and that, combined with our relatively high production volume, made overseas production a logical choice.

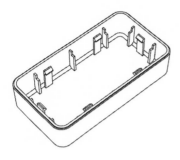

Figure 3.7

Item:	Electronics housing (bottom half)
Process:	Injection Molding
Material:	ABS plastic
Size:	2.5" (64mm) long
Finish:	Painted
Tooling:	$9900 USD
Quantity:	1000
Piece Price:	$0.79 USD
Country:	China

This part was produced using a full-hardened steel, two-cavity tool. The multiple cavities result in a relatively low part price, in spite of ABS plastic being over one-and-a-half times as expensive as polypropylene. Higher production volumes would have lowered the individual price even more. In contrast to the eyeglass frames described in the previous project, this housing was painted, which also contributed to the part cost. Injection molded parts do not require painting, but our client desired the more uniform glossy look that painting can provide. In this case, factory workers sprayed the parts as they looped around the factory floor on a conveyor belt.

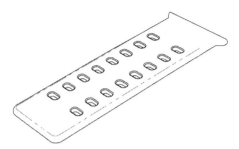

Figure 3.8

Item:	Watchband
Process:	Injection Molding
Material:	Thermoplastic Elastomer (TPE)
Size:	3.5" (90mm) long
Finish:	None
Tooling:	$5470 USD multi-cavity mold
Quantity:	2000
Piece Price:	$1.10 USD
Country:	China

This watchband has a couple of attributes that distinguish it from the other injection molding examples. One, the material is a specially engineered flexible thermoplastic that comes with a premium price tag. Two, the flexible material is injection molded over a hard plastic insert. The hard plastic insert might be purchased as a standard part, or be produced with its own associated tooling cost and individual part cost. Tooling and piece price can vary widely from vendor to vendor and the price does not necessarily correspond to the quality of the final parts. For this reason, it is important to get multiple quotes for any project and to evaluate the quality of reference parts that any given company has previously produced.

Figure 3.9

Item:	Diagnostic Meter Housing
Process:	Extrusion
Material:	Aluminum
Size:	4.5" (115mm) long
Finish:	Anodized
Tooling:	$1150 USD
Quantity:	2000
Piece Price:	$4.53 USD
Country:	USA

This part was produced locally at an aluminum extrusion facility. The typically low tooling price for extrusions can lend this process toward local manufacturing. Another contributing factor was the relatively low production quantity. At this volume, it made sense to manually cut the parts to size and to anodize them as separate steps in the process. At higher production volumes, it would be reasonable to invest in automated cutting and anodizing processes that would result in a lower part price.

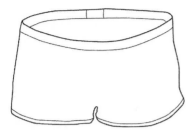

Figure 3.10

Item:	Yoga Shorts
Process:	Cut and Sew
Material:	Synthetic blend fabric
Size:	Medium
Tooling:	None
Quantity:	1000
Piece Price:	$8 USD
Country:	China

These shorts did not require any tooling, but fabric goods can often require dies for cutting the material. These were produced in China, which has a vast array of choices for cut-and-sew operations; the available choices will be significantly lower in North America, especially for a low production volume like this. Meeting minimum production requirements can sometimes be a challenge since clothing manufacturers are often accustomed to producing in very high quantities.

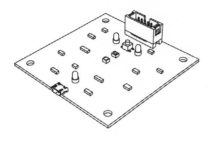

Figure 3.11

Item: PCB. 2 layers, 20 holes
Material: FR4 .062" (1.6mm) thick
Size: 3" square (76mm)
Tooling: None
Quantity: 1000
Piece Price: $21.25 USD
Country: USA and China

The price of a printed circuit board is generally comprised of the board price plus the components price plus the assembly price. The cost to assemble the components usually outweighs the cost of the bare board. This project was quoted through Sierra Circuits' online quoting tool. See the RESOURCES section at the end of the book for more information on that tool. In this case, the PCB unit price was $4.00 and the assembly price was $17.25. This PCB had fifty surface-mount components (SMT) and five through-hole components assembled to one side of the board only. Board prices can be significantly lower at higher volumes and also if production is coordinated directly with an Asian manufacturing facility. Printed circuit board manufacturing will sometimes have an associated non-recurring engineering (NRE) cost of a few hundred to a few thousand dollars, depending on the project complexity and the vendor's payment structure.

Now that we have a general idea of a variety of manufacturing processes and their corresponding tooling and part price, let's dive deeper. The next chapter begins with the undisputed favorite of volume manufacturing: injection molding. I'll discuss some design considerations that are particular to that process, and also cover a variety of other plastic production methods.

4

Plastic Manufacturing

In many ways, plastics are a designer's dream. The material lends itself to beautiful, complex shapes and to efficient production processes. While the initial tooling costs may be high, part cost is usually very low. Typical production methods for plastics include injection molding, blow molding, extruding, and thermoforming. Injection molding and blow molding tend to have higher tooling costs, while extruding and thermoforming tend to have lower tooling costs.

Injection Molding

Injection molding may be the most common plastic production process. This popular method can accommodate complicated shapes and a wide variety of features such as snaps, screw bosses and ribs that are all formed at once as molten plastic is injected into a mold under high pressure. Its low piece price and comparatively expensive tooling cost make the process most efficient at high production volumes.

Since injection molding is so well suited to high-volume manufacturing, let us have a close look at some of the specific design principles for injection-molded parts.

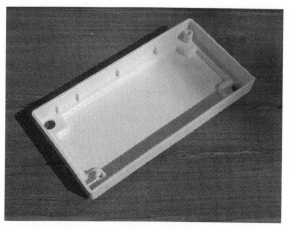

Figure 4.1 | Injection molded housing with internal ribs and screw bosses

A good analogy to this manufacturing process would be using a form to make a sandcastle at the beach. In the following figure, you'll notice that the form for the turret has angled sides that allow the sand to slip out of the mold. All "vertical" walls of a plastic part must have this angle, which is called draft.

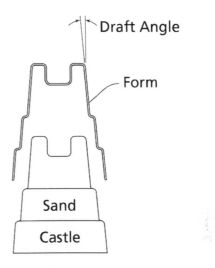

Figure 4.2

Also consider that a large mass of plastic, which is hot as it goes into the mold, will warp and shrink as it cools. To mitigate those effects, plastic parts should incorporate a uniform wall thickness that allow them to cool more evenly. For this reason, most plastic parts are cored-out, or hollow on the inside. This also goes a long way to reduce the cost since plastic is sold by weight. An efficient design will maintain the required strength while minimizing weight and cost. Our plastic sandcastle form is a good example of this. The inside shape precisely follows the curves and contours of the outside shape. The inside surface and the outside surface of this plastic piece are controlled by two mold halves called the core and the cavity, respectively.

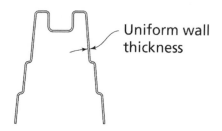

Uniform wall
thickness

Figure 4.3

Our plastic part, whatever it is, has likely ended up being some sort of cup shape at this point. If you hunt around the house and inspect some plastic things, you will see that many of them are basically a hollow cup on the inside. You'll also notice that the inside is full of little features such as ribs, hooks, and screw bosses. These features must follow special design principles in an attempt to prevent bulky, thick areas of plastic that will deform the part. If designed too thick, the little features will tend to create sink, which is a divot of imperfection on the cosmetic exterior of the part caused by non-uniform shrinking as the hot plastic cools.

Back to our sandcastle form, you'll see a new thick rib on the inside in Figure 4.4. The rib is improperly designed. In this case it should be replaced with two much thinner ribs. As it is, it adds such a localized mass of material that the outside surface will sink in as the part cools in the mold. This forms an unacceptable, and unnecessary, imperfection in our part. Thick sections like this can also cause functional problems if the sink is severe enough to cause the part to warp.

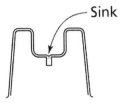

Figure 4.4

With plastic injection molding there are many competing priorities to balance. But the beauty of the process is that all this effort to get the plastic design right results in elegant, inexpensive parts. When you are producing 100,000 parts, the up-front fixed cost becomes a small component of the final piece price. For a condensed set of plastic design principles, turn to APPENDIX II, Plastic Injection Molded Part Design Guidelines. You may also download the guidelines at brightpd.com/plastic-design-guidelines.

---◇---

Tips for Plastic Part Design Success:

1 Consider draft from the moment you begin design work in CAD. Those small angles can have a distinct effect, especially on deep parts. I start with a draft angle of three degrees on external cosmetic surfaces so that the tool will be able to accommodate a medium texture. If there is reason to reduce that draft later it is easy to do, but to hack in missing drafted surfaces at the end can take a lot of rework.

2 When your design has several parts, as most do, perform an interference check in the CAD software. This check can identify any small overlapping between parts that would cause assembly problems. I've found tall components on the PCB unexpectedly interfering with the case so that I need to go back and accommodate by thinning walls or moving features.

3 Materials can be a big deal. Anything meant to be used outdoors or against the skin, for example, will need to be made with carefully selected materials. Sunlight is particularly harsh on plastics; some will hold up well, but most don't. You can also specify additives to the base plastic that will help considerably. Sweat, oils from the skin, and–worst of all–sunscreen tend to eat up plastic. So, wearable products present a material challenge. If the material must be flexible or rubbery it gets even more difficult. Start your material search early and test and prototype with different options if at all possible.

———————◇———————

Sometimes parts will require a feature that overhangs a portion of the tool and would prevent the part from being ejected from the mold. This is called an undercut. Undercuts can often be avoided by redesigning the way the part works. However, in many situations, undercuts will provide for better product function or improved assembly and are accommodated with side action in the tool. In these cases, the increase in cost and complexity of the tool is easily justified. A lifter in the tool is mechanically hinged to the main tool body and automatically swings out of the way of the undercut as the tool is opened. A slide is activated separately from the main tool motion. It can be pulled out of the way hydraulically or by some other means each time before the tool opens. Naturally, the added tooling cost of undercuts becomes more palatable as production volume increases.

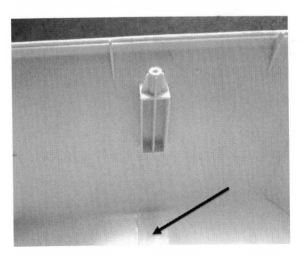

Figure 4.5 | The arrow indicates an impression left below the undercut. This is a sure sign of a lifter in the tool.

Normally, we would try to avoid designing parts with undercuts, but for the case shown in Figure 4.5, we deliberately used them to shorten some deep screw bosses. The part was about six inches (150 mm) deep and it would have required very long bosses with deep ribs. These long ribs and bosses ran the risk of showing through as imperfections on the glossy, exterior walls of the part. For these reasons, we went with shortened screw bosses accommodated with a lifter in the tool. This solution was also supported by our manufacturer as being our best option. This helped us achieve the high quality finish on the outside of our part that we were looking for.

Let's consider how to approach the undercuts in the mustache comb design. The main body of the comb is hollowed out from the end of the handle so that we can insert the little PCB. This is a necessary part of the design, but represents a substantial undercut. This hollow is achieved with a slide in the tool. At our initial modest production volume of 3000 units per year, the slide may be manually activated. When our production increases in subsequent years, we plan to invest in new, more efficient tooling, which will bring down the individual part cost. The new tools will likely be multi-cavity—that is, several parts will be produced at once, and the slide would be activated hydraulically so that it would not require human intervention with each cycle of the machine. This makes a great deal of sense, as shorter cycle times directly correspond to a lower price per part.

Blow Molding

Blow molding produces hollow objects like bottles, jugs, and tool cases. The molding equipment forces air into a tube of molten plastic, called the parison, which is blown up like a balloon into a mold. The mold controls the outer shape. The inner surface and wall thickness will vary somewhat since it is not expressly controlled; it simply follows the general shape of the outer surface. Bottles are typically blow molded from injection-molded preforms that include the threads.

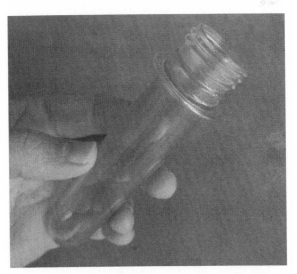

Figure 4.6 | Threaded preform for making blow molded bottles

Similar to injection molding, the plastic piece must be designed so that it will slip out of the mold. This means that vertical faces should incorporate draft and that undercuts should be minimized or eliminated.

Figure 4.7 | Blow molded gasoline jug

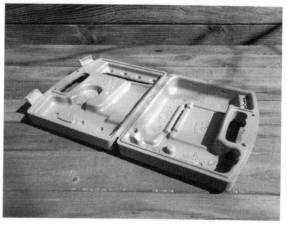

Figure 4.8 | Blow molded tool case

Extruding

Extruding produces a continuous length of material with a constant cross-section that is then cut to size for the finished part. These parts include flanges, tubes, frames, and pipes. Extruded parts often require a separate cap to finish off the end. In production, molten plastic is forced through a die and then cools as it comes through the other side. It is not necessary to consider draft for extruded parts, but the cross-section should be designed with a uniform wall thickness.

Figure 4.9 | This extruded plastic section provides shipping protection for the metal rails of this dishwasher and is meant to be discarded before installing the appliance.

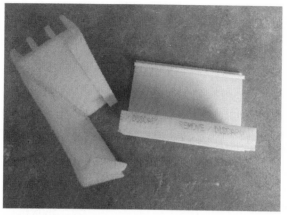

Figure 4.10 | These extruded parts provide an economical packaging solution.

———————◇———————

Case Study: Material Selection for Overmolding

Process: Injection Molding
Material: TPE, polypropylene

Figure 4.11 | Silicone rubber and TPE samples. Rubbery material selection gets complicated quickly.

We had designed a part that required two very different properties. It needed a stretchy, rubbery material and a hard, scuff resistant material. This pointed directly to an overmolded, or two-shot, process where a part is molded directly over another part. It is important to select materials that will bond. We also included mechanical interlocks in our design to make the connection as firm as possible. With all these requirements, the material selection became a real puzzle. First we contemplated whether to use thermo*plastic*, or thermo*set* materials. Thermosets such as silicone rubber can have excellent elongation and can also be formulated with great durability. We certainly could have gone this route and ended up with a quality product. However, we chose to go with thermoplastics instead because this more common process gave us more flexibility with our vendor selection. We tracked down materials and specified a polypropylene and a TPE that matched our durometer and elongation requirements. These two materials also bond well. We had a good design and good materials.

When we got first shots back, we knew we had a problem. The TPE felt wrong to the touch. When we tested it, the two parts separated pretty quickly. It turned out the factory had substituted a cheaper TPE. It had neither the mechanical properties, nor the bond compatibility that we had carefully specified. We sorted it out and ended up getting good parts. It pays off to pay close attention to materials.

◇

Thermoforming

Thermoforming generally requires less expensive tooling than injection molding, so it can make sense for lower volume production. It is also used for large parts that would be beyond the size of standard injection molding equipment. Thermoformed parts are used for items such as vending machine and kiosk panels, truck-bed liners, and packaging. The plastic clamshell containers used for strawberries are another common example.

Figure 4.12 | Thermoformed gardening tray

Figure 4.13 | Thermoformed strawberry packaging

In thermoforming, a heated sheet of plastic is draped over a pattern and formed into shape by a vacuum or other source of pressure. The part may also require secondary operations like trimming the edges or drilling holes.

In the next chapter we'll cover metal manufacturing. You will find that a couple of processing methods, particularly extruding and machining, overlap between these two material categories.

5

Metal Manufacturing

Some of the processes in this chapter are exclusive to metals. Others, like machining, are just as appropriate for a wide variety of other materials including plastics. In addition to machining we will discuss forming, casting, forging, and extruding. Machining and forming processes can often be executed with little or no tooling, although as we have seen with the examples in a previous chapter, tooling correctly applied to these methods can bring down the individual part price. The tooling costs for casting can range from relatively inexpensive in the case of sand casting to fairly expensive in the case of die casting. Forging also has a significant tooling cost because of the extreme forces that the tool must endure. Extrusions also require tooling, but the cost is often comparatively low.

Machining

Machining is often used for precise, high tolerance parts. It is also used when production volumes do not justify the higher tooling cost of forging or casting. Computer controlled machining (CNC) is automated and can often be performed with little or no tooling.

Figure 5.1 | The rectangular hollow in this sensor housing has rounded corners that correspond to the diameter of the cutting tool.

Machined parts are produced on milling machines, lathes, and other specialized equipment. In a milling machine, the part remains stationary in a vise or fixture, and the part is cut with various drill bits, milling bits, and other cutters. Milled parts will have a rounded radius in the corners of subtractive features like holes or slots. This is an artifact of the round cutting tool, and should be anticipated in the design of the part. During fabrication one or more faces of the part are hidden from the cutting tool by the vise or fixture. To gain access to all the faces requires multiple set-ups in the machine. Well-designed parts will minimize the number of required set-ups. All holes in a machined part should correspond to standard drill bit diameters if possible. See APPENDIX III for a list of those standard sizes.

A lathe is used for producing cylindrical parts. In a lathe, the part is secured in a rotating chuck. The cutting tool does not spin, but remains on a movable tool holder that can be advanced against the spinning material. Screw threads, for example, can be turned on a lathe. For a part to be fully fabricated on a lathe, it must be symmetrical about a central axis. If necessary, a lathe-turned part can be removed to a mill for additional machining.

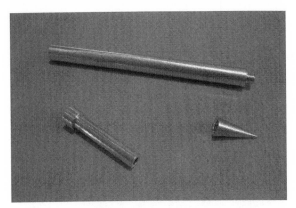

Figure 5.2 | Machined aluminum mechanical pencil by Modern Fuel. These parts were fabricated on a lathe.

Manufacturers have been employing manual machining practices since the industrial revolution, however modern CNC machining has allowed for faster, more efficient production of complicated parts. Surgical tools, industrial equipment, machinery, and specialized sports components commonly include CNC machined parts. Manufacturers take advantage of highly-automated machining processes (with high up-front tooling and NRE costs) to create consumer goods such as the Apple MacBook computer chassis.

———————◇———————

Tips for Machining Success:

1 In order to create efficiently-machined parts, consider the tools that will be used to cut those parts. Hole diameters should follow standard drill bit sizes.

2 The cutting tools for milling machines are cylindrical, so inside corners will have a radius.

3 The deeper the cut, the larger the cutting tool will have to be, so deep features should allow for a larger radius. The ratio of feature depth to tool diameter should not exceed 6:1. For example, with a .25 inch diameter cutter, feature depth should not exceed 1.5 inches. Similarly, small detailed features at the bottom of a recess can be difficult to machine. Material properties also play an important role in potential feature depth. Softer materials will allow for deeper cuts.

4 Screw threads should not exceed a depth of 2.5 times the thread diameter. For example, a ¼-20 UNC threaded hole should not have a thread depth greater than .625"

5 Deep parts that require a lot of material removal can be expensive to machine. Consider breaking parts into multiple pieces if there are isolated tall features such as posts or ledges.

6 When machining plastics down to a thin wall thickness, choose a vendor that is experienced with plastics. Special care must be taken to prevent the plastic from warping in strange ways after the machining process. Residual stresses can manifest as bent parts once much of the supporting stock material has been removed.

————◇————

Forming

Formed sheet metal starts as flat stock normally less than one-quarter of an inch thick. A flat pattern is cut using a variety of processes including sawing, laser-cutting, waterjet, stamping, drilling and punching. The simplest designs incorporate straight-line bends. These shapes can often be manufactured using standard equipment. In such designs, the flat pattern is bent into final shape using sheet metal brakes and other machinery. The particular material properties will influence how extreme of a shape and how sharp of a bend the material can take. Lower-volume production (less than a few thousand pieces) can be done with little or minimal up-front tooling, but with higher individual piece price. Higher-volume production (tens to hundreds-of-thousands) can be very efficient when the production process incorporates custom dies and other tooling. A manufacturer may also employ custom dies to form more complex shapes that are not possible with standard straight-line bending equipment.

Materials such as steel, aluminum, and brass come in different standard thicknesses, on which all designs should be based. See APPENDIX IV for standard sheet metal thicknesses. To avoid tooling costs, straight-line bends that are of a greater radius than the material thickness generally should follow standard increments: 1/8 inch increments in the US, and two millimeter increments (even increments) elsewhere.

Computer chassis, brackets, and springs are common sheet metal formed parts.

Figure 5.3 | Formed sheet metal springs

Figure 5.4 | Formed sheet metal bracket

Casting

The various types of casting include sand casting, die casting, and investment casting. Machine operators or automated processes introduce molten metal into a mold in each of these fabrication methods. Depending on the specific process used, casting can accommodate very small objects such as pieces of jewelry, and very large objects such as industrial machinery components.

Sand casting lends itself to smaller production volumes because the mold cost is relatively low. A limited number of parts may be produced from each sand mold, but then operators can easily prepare a new sand mold from a master pattern. Sand casted parts require draft and limited undercuts so that the part can be removed. Secondary machining operations such as drilling and facing are often needed to further modify sand casted parts.

Die casting resembles plastic injection molding, and is likewise suited to high volume production. Molten metal under high pressure is forced into a metal mold, often with multiple cavities, in an automated process. Die cast parts should be designed with draft and limited undercuts. The tooling for die casting can be considerably more expensive than the tooling for sand casting, but it results in parts with a good surface finish that require minimal post processing.

Unlike injection molding or die casting, investment casting can accommodate undercuts and undrafted walls, since the mold is destroyed with each part that is produced. So, investment casting lends itself to intricate geometries. In this process, a plaster cast is made from a wax core. Heaters melt the wax core out of the plaster and then the operator or the automated machinery casts molten metal into the plaster mold. Finishing processes break off the plaster cast to reveal the part inside. Investment casting tooling is relatively expensive since each part requires a new mold.

These various casting techniques are used for decorative badges, jewelry, industrial valves, and large equipment bases.

Forging

Forging incorporates ancient hand-crafting principles, though today, high-volume manufactured parts are forged with heavy, automated equipment. To forge a part, a giant hammer repeatedly slams into a hot metal ingot placed in a heavy-duty die, requiring extremely high-impact forces. After repeated blows, the metal will take the shape of the die. Forged parts often require a secondary process such as machining to finish out the shape and to detail finer features.

The heavy blows of the forging process mechanically align the grain of the metal which results in strong parts. Forged parts are commonly used for tools and equipment components.

Figure 5.5 | Forged adjustable wrench

Extruding

Metal extruding shares much in common with plastic extruding. It creates shapes with uniform cross-sections that are then cut to length. Extruded parts also often undergo secondary processes such as machining to cut out material or drill holes. The simplest extruded parts include structural shapes such as pipes, channels, and brackets. Extrusion tooling, called the die, can be relatively inexpensive.

Extrusions used in consumer goods usually require some sort of cap (often injection molded) to finish off the ends.

Figure 5.6 | iPod Shuffle, iPod Mini, and iPod Nano, with extruded aluminum construction[2]. Note the injection molded caps.

6

Fabric and Textile Manufacturing

Textile manufacturing infrastructure has become most developed in low-wage earning areas of the world. Worldwide manufacturing centers include but are not limited to China, Indonesia, and Bangladesh in the Asia Pacific, and also Brazil, Colombia, and other countries of Latin American.

Fabricators often construct initial prototypes by following hand sketches or computer-aided sketches with accompanying notes and dimensions. Once the designer approves the prototype, factory engineers develop final patterns and sewing operations from the master prototype. In many cases, it makes sense to develop a master prototype locally with a skilled cut-and-sew designer and then have the factory recreate it with processes suited to their higher volume manufacturing. It can be difficult to secure contract cut-and-sew manufacturing for low-quantity production— that is, quantities less than five or ten thousand. Check out Maker's Row (listed in the RESOURCES section) for American small batch manufacturers.

Planet Money, a podcast produced by National Public Radio, followed the materials and processes around the globe for making a printed T-shirt—from cotton grown in the United States to spinning, weaving, cutting, and sewing in various manufacturing centers, and finally shipping back to the United States. They started with a Kickstarter campaign that raised over 590,000 dollars and they produced 24,470 T-shirts. Their cost for one T-shirt ended up at $12.42. See the following figure for a cost breakdown that illustrates the various different costs associated with delivering this product.

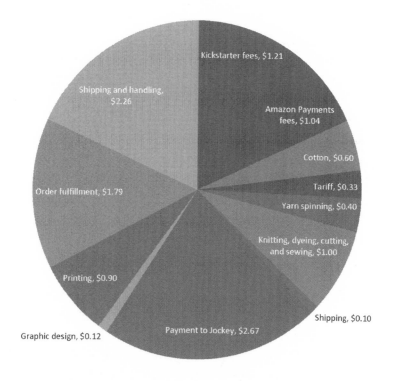

Figure 6.1 | Planet Money T-shirt Costs. Total = $12.42

PCB Manufacturing

Printed circuit boards, or PCBs, are fundamental to just about any electrical product from smartphones to refrigerators. The design and fabrication process will normally proceed in the following stages:

1. Design Concept/Specifications
2. Schematic Entry
3. Schematic, Netlist, and Bill of Materials
4. PCB Layout
5. Fabrication and Assembly Documents
6. Manufacturing

Based on the design requirements, the engineer develops a schematic that shows the interconnected design of the electrical circuit as a block diagram. The schematic, along with a bill of materials and a netlist that describes the terminals and their connections, form the parameters of the electrical design. An engineer then produces a layout based on these documents. The layout controls the exact shape and pattern of the connecting traces and components. The CAD files that control an electrical design include Gerbers, Fab drawings, NC Drill files and Pick-and-Place files. Under the supervision of a CAM engineer, automated machinery then processes the board according to the CAD files. The result is a bare board with all the appropriate soldering pads and traces. Next the individual components such as switches, LEDs, connectors, and microprocessors must be assembled to the board. A technician solders these components to the board by hand or, in high-volume manufacturing, automated equipment places and solders surface mounted components. These components are supplied in a reel, on carrier tapes, that are fed directly into the automated surface mounting machine. Some components that receive high mechanical stresses may be attached with through-holes instead of surface mounted. It might be possible to pop off surface mounted components, but through-hole mounted components would have to have their wires wrenched from the hole or broken in order to sever electrical contact. Candidates for through-hole mounting would include large elements like some capacitors that might incidentally get knocked around, or elements that take deliberate forces such as USB ports and power plugs.

Figure 7.1 | PCB layout on computer, left; and the manufactured board, right[3]

———————◇———————

Tips for PCB success:

Jessica Campbell, founder of Pristine Circuits, an electronics design services company, relates these tips for PCB design.

1 Always research your fab's capabilities before submitting your design, and make sure that your fab notes do not contradict anything in your design files. Set Design Rules within your CAD tool to keep the design within your fab's capabilities, and run Design Rule Check (DRC) before submitting your design. Some PCB fabricators offer free automated tools against which you can verify your design to their capabilities ahead of time (ex: Sierra Circuits' Better DFM Tool: https://www.protoexpress. com/betterdfm/).

2 Select components with common footprints when possible, and identify suitable alternate components ahead of time in the event of a sourcing problem. Check availability of components before selection and throughout the design process. Select components with high availability when possible. Procure risky components ahead of time. If a component does not have a pin-compatible alternative, you might consider adding a separate footprint (unpopulated) for a suitable alternate component as a backup.

3 Always request electrical testing of the bare PCBs by the PCB manufacturer. This is an automated check for continuity and short circuits on all nets. If a PCB fails electrical testing, it should not be included in the lot of PCBs that you paid for. Many PCB fabs will either include electrical testing in the cost of the PCBs, or require the customer to pay for electrical testing. Still, if it is not explicitly stated in the fabrication quote, be sure that electrical testing will be included.

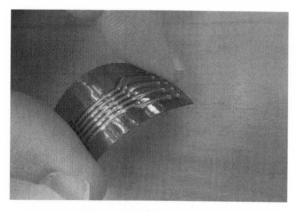

Figure 7.2 | Flex circuit

Normally, the substrate for PCBs is a rigid fiberglass material called FR4. Flexible circuits, however, can be made using a thin polyimide sheet for the substrate. Flex circuits have particular design considerations; for example, conductive traces should not have sharp corners and components may only soldered to specially prepared rigid sections. Flex circuits can be used in place of wires and cables, and the corresponding connectors can often be eliminated or greatly reduced in size. We once used flex circuits to make the connection from one earpiece of a pair of sunglasses to the other. The music player sunglasses had speakers on each side that required a connection through the hinges and across the bridge of the glasses. The flex circuit performed well, and as a result of a particular folding pattern, it survived repeated bending across the hinges.

The interior of our mustache comb offers little space for electrical components. We will be choosing the smallest reasonable switch and battery that we can, and we will specify a board thickness of .031" (.79 mm) to reduce total board height from the somewhat more common .062" (1.59 mm) thickness. The thinness of a flex circuit seems attractive, but its incompatibility with soldered-on components means it won't work for us. We would have preferred to use an off-the-shelf wireless module. Such a pre-packaged solution would handle the data transfer from the comb to the Internet. That would make the design much easier (since we wouldn't have to tackle the wireless circuits ourselves) and it would give us the advantage of the component already being FCC compliant. However, we are forced to design it from scratch so that it will fit within our limited space. We can expect to go through several additional design iterations to get the wireless component to function as it should. And as we get in to the specifics of the layout, we can expect to have to make other space-saving compromises. The board design and mechanical design will have to proceed in unison so that we have the most efficient packaging solution we can get.

8

Design For Assembly

Well-executed designs reflect an effort to reduce part count and shorten assembly time. Both of these elements influence the eventual price of the product. Where possible, a snap-hook design should be used instead of screw fasteners. Such snaps can often be included in the design of plastic parts without much of an increase to part cost or tooling cost. A bit of extra design time to include these features will be worth it to speed up the assembly process.

Figure 8.1 | The one-time snaps on this phone charger allow for speedy assembly. Note that two snaps broke when taking it apart.

Screws add to the assembly time, but may be necessary because of performance requirements or limited space. Plastic parts that are not meant to be repeatedly assembled and disassembly may use thread-forming or thread-cutting screws. Much like a wood screw, these screws make their own threads into a slightly undersized hole. Thread-forming screws are typically used with ductile plastics and thread-cutting screws are used with more brittle plastics. For more durable holes or for repeated use, a threaded metal insert may be added to the plastic part. There are similar options for sheet metal parts. Thread-cutting screws may be used in thin sheet metals, but it is more common to add threaded inserts with an automated process that forces a threaded metal lug into an undersized hole. Pem-nuts, made by Penn Engineering, come in a wide variety of sizes and are a common standard for sheet metal inserts.

Parts should be designed in a way that prevents or discourages them from being assembled incorrectly. For instance, a PCB should have orientation features so that it can be inserted into a housing only one way. Likewise, housing parts should be truly symmetrical or also have orientation features. It simply won't do to have the switches or LEDs not lining up on half of the assembled products.

The two clothespins on the right are a good example of economical part design. The two halves are connected together as one piece with a thin, flexible web of plastic:

Figure 8.2 | The two clothespins on the right are comprised of only one plastic piece.

Reducing the number of parts lowers the cost by shortening the assembly time. These clothespins could have been made even simpler by integrating a plastic spring into the design. If the whole unit were then injection molded as one piece, it could be ready to go without any further assembly.

Assemblies that rely on high tolerances should be avoided when possible, since maintaining those tolerances results in higher prices. There should be spaces or gaps between parts to compensate for regular manufacturing tolerances. Use slotted holes or other forgiving features rather than specifying tight, and costly, tolerances.

As introduced with the clothespins, a "living hinge" is a great example of a way to reduce part count and complexity. A living hinge is integrated into a single plastic part as a very thin section of plastic that can bend like a hinge. This eliminates the need for separate hardware and fasteners, and eliminates the hinge assembly step entirely. Plastic tool cases and food packaging often use a living hinge design. Polypropylene and polyethylene plastics are particularly suited to living hinges because they can be flexed repeatedly. Most plastics, especially brittle plastics, will break after bending a few times.

The dental floss holder pictured below elegantly uses two living hinges to accomplish the entire container in one piece. It folds and snaps together to contain the spool of floss and a metal cutting edge.

Figure 8.3 | Injection molded dental floss case with living hinges

Design for assembly should also include design for dis-assembly. Make it easy to break down products into their constituent components so that those individual materials may be recycled. The hinged dental floss case is a great example because it is easy to open up and separate out the metal cutting insert and the spool, which is made of a different plastic. Parts that are glued together and parts that are overmolded are more difficult to separate and less likely to be recycled. Overmolded plastics, like your toothbrush, often have a rubbery plastic joined to a hard plastic and it is usually hopeless to try to separate them. Whenever possible, plastic parts should be marked with the plastic resin type as given in the following figure:

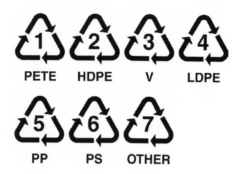

Figure 8.4 | Plastic Recycling Symbols[4]

PETE or PET: Polyethylene terephthalate
HDPE: High Density Polyethylene
V or PVC: Polyvinyl Chloride
LDPE: Low Density Polyethylene
PP: Polypropylene
PS: Polystyrene
Other: all other plastics

You may not have control over what happens to products at the end of their lives, but a well-considered design will anticipate that end and accommodate recycling or other measures to limit the heaps of trash that new products eventually become.

9

Prototyping and 3D Printing

New product development requires many, and often countless, prototypes. After all, they are the vehicle for correcting errors and improving the design before you shell out large amounts of money for tooling and ramp up your production. Many projects employ the following categories of prototypes during development:

Proof-of-Concept
Appearance Model
Working Prototype
Manufacturing Prototype

The first prototypes often begin with the designer tinkering in the garage or shop. These cobbled-together prototypes, called proof-of-concept prototypes, are often a necessary step in proving out new ideas. They may not look good, but they can provide a convincing argument for the idea.

An appearance model is a kind of prototype that represents the size, shape, colors, and design of the intended finished product, but is not functional. These models can be useful for communicating the feel of the product among various departments of the business, for testing marketing requirements, attracting investment, and they can even be a stand-in for advertisements, videos, or printed media.

A working prototype, on the other hand, proves the function of the product. In fact, this term is often interchanged with the label, functional prototype. This might be a more refined version of the proof-of-concept prototype, but it still makes little attempt at having the correct appearance, or of using correct production materials. However, it provides an efficient method for testing the engineering principles of the product. Working prototypes may go through several revisions.

Sooner or later you will need to make prototypes that reflect the as-manufactured design. These manufacturing prototypes are the result of careful engineering and the individual parts are designed to match true manufacturing processes.

Prototyping technology is rapidly evolving and you will want to choose the right tools to best represent your design. Let's look at some definitions before continuing. No doubt you are already familiar with 3D printing:

3D Printing

Any computer-controlled additive process used to build three-dimensional objects. In this process, a file is sent from a computer to the machine, and instead of printing on paper, a machine will build up a real object complete with height, depth, and width. There are many different processes and materials available. The method and material should be carefully selected to best match the part requirements.

Additive Process

A fabrication method that adds elements together to achieve the final result. 3D printing is an additive process. These machines apply very thin layers of material on top of each other to build up depth until it achieves the full part. As an analogy, consider a log cabin or a brick house as the result of an additive process. Individual bricks or logs are added to achieve the final result, a house.

Subtractive Process

A fabrication method that removes material to achieve the final result. For instance, a machined bracket represents a subtractive process. The process starts with a solid block of metal and then material is cut and drilled out to achieve the final shape. Subtractive processes have been used for ages by woodworkers and metal workers. In the modern age, subtractive processes have been greatly streamlined with CNC machining.

Figure 9.1 | The pile of chips created as I machined a groove in this track is testament to machining's subtractive nature.

Rapid Prototyping

An automated process for producing prototypes that includes 3D printing, CNC machining, and other computer-controlled fabrication methods.

Additive Manufacturing

This is 3D printing performed on an industrial scale. Additive manufacturing describes the 3D printing processes that are used to fabricate fixtures, tooling, and production parts. Imagine larger, more expensive machines that produce durable parts in metal or plastic. These processes are often combined with traditional fabrication methods such as machining to produce the final part.

Rapid Manufacturing

This is an inclusive term that encompasses all of these computer-controlled processes, whether additive, or subtractive, for prototypes, or volume manufacturing.

Now let's study some specific rapid manufacturing processes. The following chart compares selection criteria that will help in determining the best process for your application.

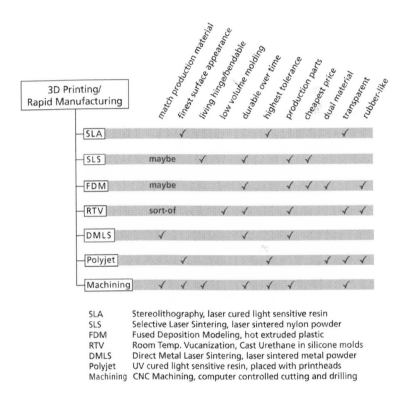

SLA   Stereolithography, laser cured light sensitive resin
SLS   Selective Laser Sintering, laser sintered nylon powder
FDM   Fused Deposition Modeling, hot extruded plastic
RTV   Room Temp. Vucanization, Cast Urethane in silicone molds
DMLS   Direct Metal Laser Sintering, laser sintered metal powder
Polyjet   UV cured light sensitive resin, placed with printheads
Machining   CNC Machining, computer controlled cutting and drilling

Figure 9.2 | 3D printing matrix. Chart download available at brightpd.com/3d-printing/

SLA

The SLA (stereolithography) process begins with a vat of liquid, light-curable resin. As with most additive processes, the part is built in very thin layers, one layer at a time. A laser draws out the layer cross-section and hardens the liquid wherever it passes. Then the part sinks just a tiny bit into the liquid, and a new layer cross-section is drawn out. The SLA process is known for its ability to create fine-detailed features. However, the materials, by nature of their light-sensitive properties, will degrade over time with exposure to light and humidity. After a period of months SLA parts can tend to become brittle and or deform, especially if left exposed to light.

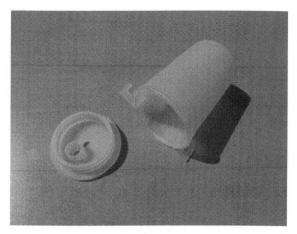

Figure 9.3 | SLA sensor housing, material: DSM Somos 9120

———◇———

Case Study: Using SLA to Approximate Frosted Plastic

Process: SLA
Material: DSM Somos 9120

We were working on a project that had a translucent plastic bezel. A main visual element of the design featured an LED that shined through with a nice frosted glowing effect. We were exploring different options for making prototypes. Some earlier prototypes had been made using a cast urethane process. With careful specification of the material and surface finish, they came out pretty good, but at a price. That process would have been better-suited if we were making fifteen or twenty parts, but we just needed a few. We ended up building the part by SLA out of a polypropylene-like material, DSM Somos 9120. The material matched the frosted translucent look we wanted very well.

———◇———

SLS

The SLS (selective laser sintering) process builds parts from finely powdered plastic. The most common material is nylon, which is also available in glass-filled and fire-retardant formulations. Several other plastics are also available. This process also uses a laser to draw out layer cross-sections, this time in a bed of powdered plastic. The laser sinters, or melts and fuses, the plastic along its path. These parts have a similar durability over time as injection molded plastic parts. In fact, the SLS process is increasingly being used for production parts. The resolution and surface appearance is generally not as fine as SLA.

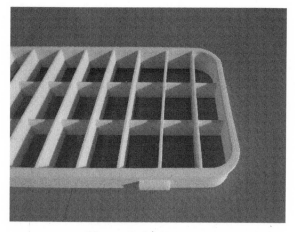

Figure 9.4 | SLS vent

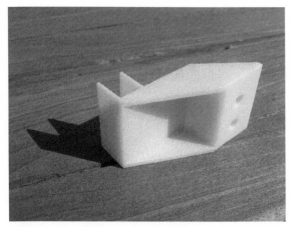

Figure 9.5 | SLS bracket

---◇---

Case Study[5]: Rapid, Living Hinges with 3D Printing

Application: Manufacturing prototype for wearable consumer product
Process: SLS
Material: Nylon 11 EX

Reserve Strap contracted Bright Product Development to design a watchband with integrated batteries and electronics that extends the battery life of a smart watch while being worn. An articulating frame with six living hinges between the electronics compartments provided the necessary flexibility to wrap the band around a user's wrist. We produced a manufacturing prototype by SLS with Nylon 11 EX material, which provided a very similar function to the production part.

Figure 9.6 | SLS 3D printed prototype

The prototype design was slightly different than the injection molded part design since the SLS process we used required a minimum build thickness of about .025". This posed a challenge because the final part design for injection molding called for living hinges which were slightly thinner than this minimum SLS build thickness. To resolve this, we manually filed down the 3D printed hinges down to the final part dimensions.

Figure 9.7 | The final injection molded
production part

The SLS process is suited to living hinges
since nylon is one of the few plastics that can
withstand repeated bending (although not as
well as polypropylene or polyethylene.) We did
file down the wall thickness in the hinge area
after printing it, but we later discovered that
the minimum SLS build thickness provided
for adequately functioning hinges without any
further modification.

◇

FDM

The FDM (fused deposition modeling) process is responsible for the surge in 3D printing popularity. Many of these processes have been used in industry for decades, but with the advent of new inexpensive FDM machines it has become more widely accessible. Machines used to cost in the range of 100,000 dollars but desktop versions are now available for under 200 dollars.

The FDM process extrudes a thin filament of plastic through a hot nozzle in much the same way as hot glue in a glue gun. Then, the machine draws out a bead of plastic in the shape of the layer cross-section. It draws out layers on top of each other to complete the part.

Many plastic materials are available for this process, and the material cost is the lowest among 3D printing technologies. Rolls of plastic filament may be ten times cheaper than an equivalent quantity of SLA resin. Many newer machines accept any filament rolls with the correct diameter; however, some manufacturers limit usage to their own, more expensive filament cartridges.

FDM can produce parts with similar detail and durability as SLS parts. However, the layers are more visible than the layers in SLA or SLS parts.

Figure 9.8 | FDM support material detail (bottom) after it is cracked off of the part (top).

---◇---

Case Study: Make Your Own Cordless Drill

Process: FDM

This is a project that I use in a CAD/Plastic Design class that I teach. It brings together some advanced 3D CAD techniques and also requires a good understanding of plastic design principles. You can cannibalize the battery and all the working parts of a cheap cordless drill and design your own custom housing. Because of the constraints of how all the parts fit together, you'll need to use good plastic design principles such as a uniform wall thickness and good screw boss design. With some careful CAD design and 3D printing using FDM, you'll have your own custom working drill and a great portfolio piece.

Figure 9.9 | Drill components

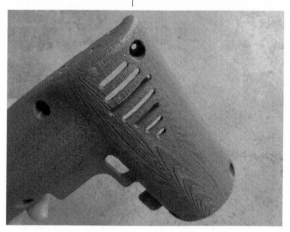

Figure 9.10 | FDM printed housing

RTV

RTV (room temperature vulcanization), also called cast urethane, involves a multistep process that would not strictly fall under the 3D printing category. This technology provides a bridge between processes that produce one part and higher volume production processes. First, one part is produced by any rapid manufacturing method (most often by SLA or CNC machining). Then, a soft silicone rubber mold is made by pouring it over the positive pattern. The pattern is removed, and finally, a liquid urethane plastic is poured into the mold, where it hardens after a short amount of time. About twenty parts, depending on part complexity, can be cast in the soft silicone mold before the mold begins to deteriorate. The RTV process can produce parts with a high-quality surface finish and a great likeness to injection molded parts. If multiple examples of a prototype part need to be produced, this can be an economical choice. Also, this is the best method for producing durable rubber-like parts. Other 3D printed rubber parts tend to tear or crumble.

Figure 9.11 | Flexible RTV watchband

DMLS

The DMLS (direct metal laser sintering) process resembles the SLS process, but with metal powder instead of plastic. Powerful lasers sinter the metal powder to build the part up, layer by layer. A variety of metal powders such as stainless steel and titanium are available to use with this technology. Engineers often harness this process for manufacturing parts that would be difficult or impossible to produce by traditional machining or casting methods. And DMLS has the added benefit, in contrast to casting, of not requiring tooling. However, if it is possible to do, machining can often be a less expensive fabrication method. Example parts include fans, manifolds and ducting. Automotive and aerospace manufacturers have begun to use this process to make engine parts.

Polyjet

The polyjet process is based on inkjet printer technology. Imagine sending a paper though the printer many times until the ink builds up a certain depth. This is analogous to the polyjet process, except that instead of ink, the printer uses a light sensitive resin similar SLA resins. A layer of resin is delivered by printheads to a build plate, cured by UV light, and then a new layer is built on top of the previous layer. This process accommodates multiple materials and colors. Overmolded "rubber", for example, can be printed with the hard plastic part. Or, a full color part can be printed. This process can produce fine detail and a high quality surface similar to SLA. This process tends to be more expensive than FDM or SLS.

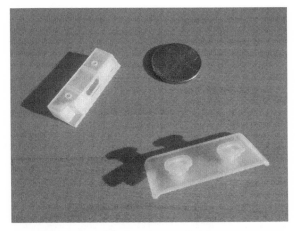

Figure 9.12 | Translucent Polyjet parts

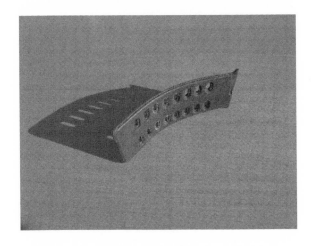

Figure 9.13 | Flexible polyjet band

———————◇———————

Case Study: 3D Printing Price Comparison

Processes: Polyjet, SLS, and SLA

We designed an adapter for a small appliance. The production part was to be injection molded, and we needed some prototypes to test the function and appearance before going to tooling. We sent the CAD out for quotes with several different 3D printing processes in mind. An SLS part would provide the most durable prototype and for that reason was the best stand-in for an actual production part. So, we made one by SLS. However, we also needed an appearance model. SLS parts always tend to retain their rough sandy look and paint does nothing to improve the matter. So, we also quoted SLA and polyjet. Both of these can be sanded and painted to a very nice surface finish. However, in this particular case, I was concerned that the screw bosses on an SLA or polyjet part would crack (and it turns out they did.)

Figure 9.14 | Adapters. The SLS prototype is on top, and the polyjet prototype is on bottom.

We ended up going with polyjet because we were able to print it in a grey color that closely matched the rest of the product. A sanded and painted finish in either polyjet or SLA would have looked better, but would have also cost more. 3D print build prices are usually related to the material volume. This part was basically a hollow shape, 13.5 x .75 x 4 inches and with .079 thick walls (342 x 19 x 100 mm, 2mm thick.) The quotes varied considerably, by both vendor and process:

Polyjet	$572
SLS	$220
SLA, vendor 1	$420
SLA, vendor 2	$189

News reporting these days makes it seem like 3D printing is inexpensive. It can be, especially if you do it yourself on your own machine. And prices will probably drop more as the industry becomes commoditized. But for now, high quality prints from professional service bureaus do not come cheap.

———————◇———————

CNC Machining

CNC (computer numerically controlled) machining is an automated subtractive process that can be as quick as 3D printing when tackled by experienced professionals. An unlimited variety of plastics and metals can be machined. The process is suited to prototypes and to volume manufacturing under the right circumstances, such as when very precise parts are required at a moderate volume. It can have fine detail and good surface finish.

Figure 9.15 | CNC machined PCB housing

———————◇———————

Tips for 3D Printing Success:

1 It is important that the CAD files for 3D printed parts are "watertight". That is, they cannot have any tiny holes or gaps. The junction of complicated surfaces can sometimes have this kind of flaw, and it won't be immediately apparent. Some surfacing design programs such as Maya, Rhino, and 3D Studio Max allow models with gaps and holes. These will have to be closed up in order to make a successful print. The software should have a tool to check the integrity of the surfaces. Little tags or dangling surfaces also cause problems with 3D prints. Any of these infinitely thin surfaces cannot be built in the real world and will cause the print to fail or even prevent it from starting.

2 Three dimensional printers build from STL files. To prepare your CAD for a print, save or export individual parts as an STL. This format tessellates the solid file into tiny polygons. Adjust the accuracy just fine enough to reproduce the surfaces. An overly fine mesh does nothing to improve surface quality and can vastly increase the file size, which can also have a detrimental effect on build speed.

3 To prepare your file for CNC machining, save or export in the .STP (STEP) format. Most equipment can be programmed from this generic format.

———————◇———————

Producing 3D Printed Parts

A service bureau provides easy access to 3D printing technology. You send them the CAD files and they mail you the parts a few days later. An advantage of using a service bureau is that you get to choose any material and process that best fits your current need. Or, you could purchase a 3D printer and print in-house. Keep in mind, industrial 3D printers can range from tens-of-thousands to well over a 100,000 dollars. Desktop FDM and SLA machines can range from a few hundred to a few thousand dollars. Make Magazine has an online 3D printer buyer's guide that is worth studying. Another option would be to source parts through a 3D printing marketplace. These marketplaces facilitate transactions with local people and businesses that have 3D printers. The prices tend to be lower than a dedicated service bureau, although the parts can be of unpredictable quality. See the RESOURCES section at the end of the book for more information on these options.

Some of the magic of 3D printing starts to dissipate when you realize the effort it takes to produce nice parts. Builds sometimes fail, or just come out funky; in such cases they have to be redone. And then, there's the post-processing. Most 3D printing methods do not produce parts that are ready to use straight out of the machine. The parts usually require some specialized methods to clean off support material or improve the surface finish. This could include bead-blasting, water-blasting, chemical baths, manual support removal, sanding, and many others processes. The post-processing time can range from minutes to hours, and it is easy to underestimate. In fact, some bureaus specialize solely on the post-processing of 3D printed parts as a service for additive manufacturing facilities.

Figure 9.16 | A 3D printed part produced on a Form Labs (SLA) printer. The lattice structure is the support material and will be removed.

Figure 9.17 | The dots on the part are a vestige of the support material and will be sanded off.

Just as prototypes go through several stages, the factory will similarly progress through various builds on the way to mass production. We'll look at that progression in the next chapter.

10

From Prototype to Mass Production

Manufacturing Validation

When designing products for a well-validated market and when the production methods are within conservative limits of traditional manufacturing (does this ever happen?) it can be practical to go directly to high-volume manufacturing after the prototyping stage. Often, however, a product will go through additional rounds of manufacturing validation before ramping up to full production. There are several cases where this will be an important step in the development cycle:

1. Current product demand is low and can be met with less efficient (and cheaper) tooling. New tooling can be purchased as market demand increases.

2. Market demand is unknown, and is tested with less efficient (and cheaper) tooling.

3. The product requires specialized tooling or processes. In other words, the parts are difficult to make. One example would be wearable electronics products. Bending, stretching,

and the extreme shapes of these products often require specialized production techniques.

For plastic parts, "soft" tooling, can often address low production requirements with a cheaper tooling price. The life of the tool is low, that is, it can produce fewer good parts before it wears out, and individual part price may be high. Soft tooling is often made out of aluminum instead of hardened steel ("hard" tooling) and automated lifters or hydraulic actions may be replaced with manual processes. For instance, an undercut in the part may require the operator to manually pick out a piece of the tool instead of having that action incorporated into the movement of the tool itself.

Sometimes soft tooling is used as a bridge on the way to higher production volumes. The tooling is purchased with the knowledge that it will soon be replaced with higher capacity and more efficient hard tooling. In these cases, it is often called bridge tooling.

Soft tooling can also be used as a prototyping method. Proto Labs, for instance, advertises "25 to 10,000+ parts shipped in 1 to 15 days" with injection molded soft tooling. Of course, such efficiencies come with some restrictions on part geometry and size. Nevertheless, some form of soft tooling is sometimes necessary to determine the manufacturability of complicated parts.

In addition to aluminum soft tooling, prototype tooling can now also be built by 3D printing. Remarkably, it is possible to injection mold plastic parts into these 3D printed plastic tools. A leading process builds the molds by polyjet. These mold halves can be attached directly to the molding machine or set within a metal mold base. Quantities in the twenty-five to fifty range can be produced before the mold begins to deteriorate. Speed is the major advantage with this method, since the mold tools can be produced in about a day. Cost is also lower than many other tooling methods, but the specialized polyjet materials do not come cheap—a pair of 3D printed mold halves for a hand-held part can easily cost over a thousand dollars.

───────────◇───────────

Case Study: Prototyping with 3D Printed Tools

Process: Injection Molding
Material: TPE, Thermoplastic Elastomer

We were having trouble prototyping a soft rubber-like part. We would have used a cast urethane process, but the service bureaus wouldn't do it because the part had a long hollow inside that wouldn't work with the silicone molds. We ended up injection molding into 3D printed polyjet tools. Stratasys recommends Digtal ABS (RGD5160) as the longest lasting polyjet mold material, but we built with Vero White because the print time was faster.

Figure 10.1 | Polyjet molds coming out of the 3D printer

In addition to core and cavity mold halves, we printed a long insert that provided for the hollow portion of the part. The insert failed on the first try. So we machined an aluminum insert to use with the polyjet mold halves. That worked, but it also began to deform after a few shots. Finally, we went to a machined steel insert, which worked the best. We never did get perfect parts, but they were close enough to evaluate the design. It is truly a marvel to get injection molded parts this fast.

Figure 10.2 | Polyjet tool in the injection molding machine

The great thing about this process is that it also allowed us to test production materials. We dialed in the material requirements from these tests. We discovered that all the readily available TPE blends we could find were too stiff for our application, so we worked with PolyOne to custom blend a material. One thing that we learned about TPEs is that it is very difficult to formulate a soft material that does not feel tacky to the touch, while also maintaining good environmental resistance. We probably got fifty good shots out of the tools before deformations started causing problems.

———————◇———————

Even if a factory is producing a product using conservative and established fabrication techniques, it will normally step through several builds on the way to mass production. This allows them to refine their processes, to make adjustments, and to order and stock sufficient material. These builds include:

Engineering Build, EB
Manufacturing Build, MB
Pilot Production, PP
Mass Production, MP

There is some overlap here with prototyping on the engineering build; the definitions are not fast across the industry. The key point is that the factory has access to this build for evaluation purposes, while they may not have seen previous prototyping efforts. The engineering build will give the factory an idea of the required processes and an indication of the complexity of the product. On a scale of completeness, the engineering build would fall somewhere after the manufacturing prototype. This build may come before tooling with the parts being made using any variety of prototyping methods. Or, in an ideal scenario, it would come after tooling from the very first shots from the production machines. The parts need not be the right color, or meet the final appearance requirements, but rather should provide the form, fit, and function of the final production parts. Engineers hand-assemble several examples of the product off-line from the factory floor. The engineering build may also serve for regulatory testing if the product needs FCC, UL, CE, or any other certifications.

The manufacturing build integrates the production parts into the factory assembly process. At this time, injection molding tooling is finalized and technicians apply appropriate textures and finishes to the molds. This step proves out the full manufacturing and assembly, and provides opportunity for process adjustments as necessary. The manufacturing build may be produced in some quantity less than one hundred and can be used for internal evaluation and testing.

There will likely be a gap of several weeks between the manufacturing build and pilot production. During this time the factory will order and stock the materials required for higher volume production. During pilot production, the factory implements whatever they learned during the manufacturing build and it further fine-tunes the manufacturing process. The factory prepares for the volume and schedule requirements of mass production with this build.

Finally, the factory is ready for mass production. They will continue to adjust their processes during production as they seek and discover additional opportunities for greater efficiency. If you have made it to mass production, you have reached a real milestone that represents the culmination of many months of hard work (along with a sizable monetary investment!) Well done. Now continues the equally hard work of business building and sales.

In the next chapter we'll step back and look at the engineering tools and documentation that supports your product. This documentation provides a complete description of everything going into the product and provides a basis for an RFQ (Request for Quote) and ultimately provides instruction for the manufacturing itself.

11

Documentation: Drawings and 3D CAD

For most any type of manufacturing you will design and engineer parts using 3D CAD software. The 3D CAD models, together with 2D engineering drawings, provide the necessary information for producing prototypes and final production parts. At a minimum, the 2D drawings should specify material, finish, and critical dimensions with acceptable tolerances. The 2D drawings serve as a specification between you and the manufacturer of what, precisely, they agree to deliver. A carefully specified Bill of Materials (BOM) is part of the engineering documentation and should accompany the 2D assembly drawings.

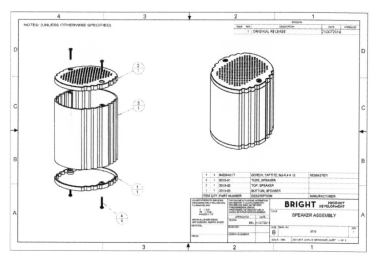

Figure 11.1 | Assembly drawing with BOM

These important documents will reduce the chance for misunderstandings with your manufacturer that could otherwise lead to scrapped parts and wasted money. This engineering documentation also serves as the basic element of the Request for Quote, RFQ, when approaching potential vendors.

Any manufacturing process has a related range of tolerance to which it can reliably produce parts. In general, the tighter the tolerance you specify, the more expensive it will be to produce the part. So, it is necessary to allow large tolerances wherever possible and tight tolerances only where it is important to the function of your product. Careful tolerances are necessary, for example, where a pin presses into a hole, where one piece fits inside of another, and generally where any two parts come together in an assembly.

Let us consider an example: a plastic door requires a snug fit so that the door will click shut to the housing. Most of the dimensions are not critical and we are able to accept standard plastic injection molding tolerances. But, we give a carefully considered tolerance on the spacing between the snaps. The tolerance, designated by a +/- range, gives a band of acceptability to the actual manufactured part. We also give overall reference dimensions, designated in parenthesis, for informational purposes only. This part will be manufactured from the 3D CAD file along with the additional information and tolerances given in the 2D drawing. See the following figures:

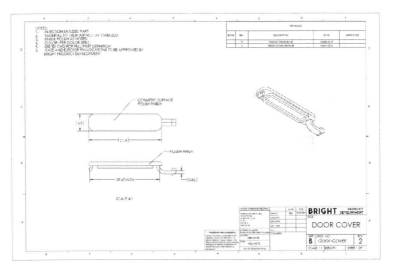

Figure 11.2 | The drawing has a note that refers to the 3D CAD.

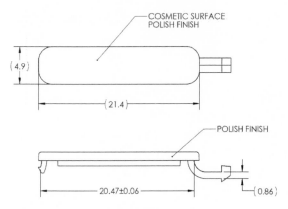

Figure 11.3 | Reference dimensions (in parenthesis) and critical dimension with tolerance

Now, another example: The following machined part includes many dimensions on the drawing, but it still relies on the 3D CAD for full part definition. For this part we call out dimensions with tolerances for an O-ring groove that gives a water-tight seal. The truth is, all other dimensions on the drawing also have an associated tolerance, even though it is not stated next to the dimension. All dimensions on the drawing follow the tolerance chart given in the title block unless they have their own explicit tolerance. This drawing also calls out the threaded holes.

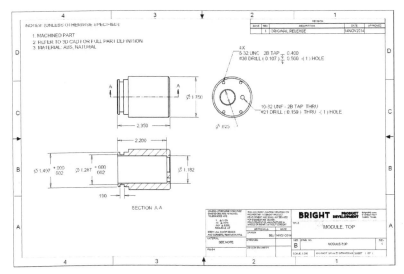

Figure 11.4 | The drawing calls out critical dimensions and threaded holes.

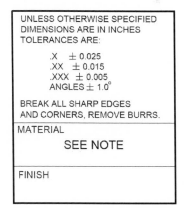

Figure 11.5 | Tolerance Chart, excerpted from the title block

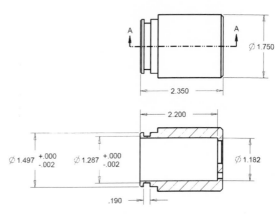

Figure 11.6 | Tolerances that vary from the chart are stated explicitly next to the dimension with a +/- designation.

Some parts will be fully dimensioned in the 2D drawing so that the 3D CAD is not necessary for manufacturing. The following drawing of a fully dimensioned sheet metal part, for instance, contains everything needed to make it.

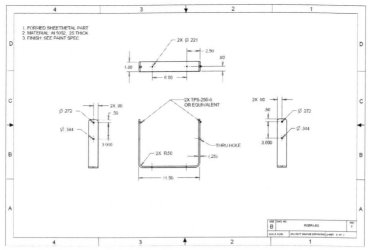

Figure 11.7 | Fully dimensioned drawing of a sheet metal support riser

Model Based Definition, or MBD, provides an alternative to 2D drawings. Many of the leading software packages allow tolerances, notes, and other data to be attached directly to the 3D CAD model. In this way, the important auxiliary information that has traditionally been captured in a drawing can be integrated within the 3D CAD file itself. Businesses that are able to control all steps of manufacturing within their organization are currently most likely to be able to take advantage of Model Based Definition. Aerospace and automotive industries have been early adopters of this technology.

To create the drawings and 3D CAD, the parts need to be built with 3D modeling software. Some of the more common programs in the industry include Creo Parametric, Solidworks, and Inventor. Each of these programs has the tools to create well-engineered parts. These programs are history-based and parametric. That is, each feature you add to the model is recorded in a time-line. You may go back and adjust dimensions and modify features, and then those changes will propagate through the model. In other words, the modeling process is arranged so that ongoing adjustment and change are possible. CAD software is a powerful tool in the hands of an experienced engineer.

For the maker, there are some free/low-cost CAD options available that can be used to create adequate engineering models, but they do have their limits. The acquisition of good CAD skills is a continual pursuit. It is a craft and a nurtured skill that takes time and practice. The landscape of CAD software, of course, is constantly evolving. The following chart outlines some of the current CAD options.

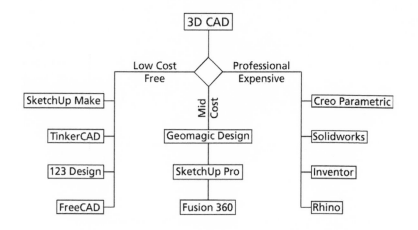

Figure 11.8 | Engineering CAD landscape

12

Vendor Selection: Where in the World to Manufacture?

As late as the 90s and into the 2000s, the conventional wisdom was that to have a competitive product with competitive pricing, it must be produced in China or in other developing areas of the region. For many products this is still the case. Why? Price, and just as importantly, manufacturing infrastructure. Everything you need to build a bicycle you can find within a small radius on the island of Taiwan for a good price. You'll have tougher luck finding that in North America. Need a specialty screw for your electronic device? If you are in Shenzhen, China, there's probably a plant around the corner that will make it for you. There are massive ecosystems of manufacturing in Asia that make it possible to do things that would be difficult to accomplish elsewhere.

That said, wages (and prices) in China are rising and communication difficulties can delay schedules and have other costly consequences. A client described this well. We had designed some small pieces of furniture for a company called Brixton Banks. I had a conversation with the owner, Amber Hagopian, about her manufacturing experience here and abroad.

BBJ: "I've found that increasing the opportunities for effective communication usually increases the chances for successful product fabrication and delivery. That's one reason fabricating close to home, versus overseas, can be an effective option. Do you have any comments about this that you can relate specifically to Brixton Banks products?"

Hagopian: "I would definitely agree with this. Initially, it seemed like we could just send over a purchase order with some drawings to a factory, but what we found after sourcing in China and then back in the USA was that there was very valuable input on design and materials from the USA based fabricators that we did not get from the factories in China. This was due to several reasons, one being that the USA based manufactures were more willing to work with a new company and provide their input and secondly there was still a language barrier even with English-speaking coordinators overseas. Our products have gone through several design tweaks based on recommendations from our states-based manufacturers, which has made our products better quality, lighter in weight and more cost effective."

BBJ: "What were your biggest hurdles related to manufacturing your first product in China?"

Hagopian: "The biggest hurdles we faced fabricating a new product overseas was reaching the minimum order quantity and feeling confident the factory we chose was going to provide what we ordered and get it to us on time and in good condition. We did not use a USA-based third party company to choose a factory and ended up with a 20 foot container of product that we ended up paying to have disposed of because there was no quality control in place. The materials used for actual production were much cheaper quality than the prototype the factory originally provided and we were unable to sell the items. We did not have proper agreements or warranties in place and ended up taking a total loss with no recourse against the Shenzhen-based company. In addition to taking the loss, we ended up without inventory to supply the demand we created for our product, which has been an ongoing issue. For one product we cannot get fabricated cost-effectively in the USA, we are now using a states-based company that sources factories overseas, provides quality control and coordinates shipping to our distribution center."

BBJ: "What were the biggest advantages?"

Hagopian: "The only advantage we have to sourcing overseas is costs. Even with shipping cost, materials and labor are so much cheaper over there that if we hadn't initially sourced overseas we would not have been able to bring the product to market because there was not enough margin to make it worthwhile."

BBJ: "How would you compare/contrast that to your manufacturing experience with the current product in the US?"

Hagopian: "Here in the states we have been able to order smaller quantities of product as we tweak our design. This also helps with cash flow, as we do not have to place large minimum order quantities and then pay to store them as we ramp up our business. Additionally, the USA-based manufacturers have been very helpful in perfecting the product design for aesthetic and cost. Lead time for smaller quantities is much less in the USA, and being able to be in communication over the telephone with the fabricators here has been helpful."

Manufacturing close to home can have a variety of advantages. For one, a shorter supply chain can allow you to manufacture in smaller batches. Orders from China require large volumes of merchandise that remain static as they are slowly sold throughout the year. Smaller production cycles allow more frequent product innovation, and smaller upfront purchases. Furthermore, it is easier to fix any design problems without risking large faulty production runs. Local manufacturing is often much faster. Overseas communication frequently lags, and even urgent shipments can be delayed. And an added benefit is that there are fewer chances for leaking intellectual property.

Decision time: Where Do We Manufacture?

Have we answered the question yet, of where to manufacture? Not quite, but let's distill the discussion into some categories that will allow you to make an informed decision.

1. Production quantity in the hundreds or thousands can tend toward local or regional manufacturing. Production volumes in the tens-of-thousands and greater can tend toward worldwide manufacturing centers.

2. Regional manufacturing is advantageous for rapidly-changing products and short development cycles. It is faster. New technology fields can also thrive with regional manufacturing. Think 3D printing and drones.

3. Local manufacturing can be less risky, especially for those inexperienced with bringing a product to market.

4. Some production locations will be dictated by where in the world the expertise, factories, and supplies are located. Garments, bicycles, and electronics are examples of this.

5. When manufacturing overseas, please retain the services of a sourcing/supply chain/quality agent. It will be money well spent.

───────◇───────

Tips for Selecting PCB Vendors:

Jessica Campbell has this to say about selecting PCB fabrication vendors:

"I always stick to qualified domestic vendors during the early design and prototyping phases. Domestic vendors can produce high quality PCBs in a shorter time frame, with lower risk of errors. When issues come up, they can be fixed quickly, and a visit to the factory floor is easy. The only real benefit of manufacturing overseas is cost, as PCB fabrication overseas can be up to fifty percent cheaper than in the US. My recommendation is to stick with domestic vendors during the early design and prototyping phases, and move to overseas vendors once the design is polished and production is ramping up.

Good communication between the PCB designer and the CAM (Computer Aided Manufacturing) engineer before PCB fabrication is especially important. A good CAM engineer will review the design files, identify any questions or discrepancies that could compromise the quality or functionality of the PCB, and communicate these discrepancies to the PCB designer with proposed solutions for each. When the PCB designer has approved or otherwise commented on all proposed changes, PCB fabrication will move forward.

Next you can move production to your favorite contract manufacturer who handles PCB fabrication and assembly for larger production volumes. My observation has been that most of the contract manufacturers have relationships with overseas PCB fabs and simply act as a "middle man" for sourcing the bare PCBs. They do charge a premium, but they offer advantages, such as supply chain management, which can save you money in the long run."

◇

We have decided on a hybrid approach for our mustache comb. For our initial production run, we will work with an established stateside PCB manufacturer that will coordinate the production in China. This is more expensive than going to a Chinese factory directly, but it will be easier to manage, it will be faster than finding and vetting a Chinese factory, and we will be better assured of ending up with a quality product. We plan for a second phase of manufacturing to accommodate increased sales and at that time we can transition to our own Chinese contract PCB manufacturer. During this time we will also be working on other cost reduction measures.

For the mechanical components, we are exploring three options. First, we found injection molders that make plastic combs. Surely there are hundreds of factories that make combs, but we could only find three that do. After many emails and some calls, we were convinced that one of them was also experienced with the kind of overmolding that we wanted to do. We also followed a second track of injection molders recommended by personal contacts and referrals. None of these had experience with combs, but several had experience with personal care products and electronics. We chose to follow up with one of these recommended Chinese factories that had proven themselves with timely and effective communication. Our third option was a lesser known stateside rapid injection molder that we found through dedicated internet searches. Their prices and claimed production time were impressive.

Vendor quoting: The RFQ

You have done your best to narrow the pool of potential contract manufacturers to two or three vendors. These vendors have appropriate experience with your product category and are a match for your potential production volume. Now you can approach them with a Request for Quote, RFQ. The RFQ is comprised of product specifications, engineering documentation, forecasted production volume, and your company information or investor slides. This comprehensive package represents a starting point with your manufacturer and will require an investment of time from them. They will evaluate their continued profit potential based, in part, on the health of your company, sales volumes on previous products, and market position, so in a sense you are also selling your success as a company to them. The forecasted volume should include estimates for initial production volume, volume per month thereafter, and total volume for the life of the product. Together with the engineering documentation and specifications, they should be able to produce a realistic quote. The quote will come with order minimums, stated either as Minimum Order Quantity, MOQ, or Minimum Order, MO, in dollars. As you move forward with a vendor, the RFQ documentation will be revised based on the production processes and recommendations of the contract manufacturer.

Controlling your supply chain

We have lost much of the meaning of the words, "quality control". It sounds like a good idea, and it is. It also sounds like a system that I would want to put in place once the company got big enough to funnel resources towards it. Then we would get our guy with a clipboard checklist on the factory floor. This misses the whole point.

Quality control is, simply, control of the supply chain. This includes control of each of the individual purchased components, custom manufactured parts, assembly processes, and so on. It begins with the very act of vendor selection. What does this mean for a business with limited resources? First, this means vetting all of your suppliers and manufacturers. Do your key suppliers care that you exist? Are they responsive to questions over the phone and do they address your concerns? Can you talk to the boss of your contract manufacturer and are you assured that the company is committed to success of the project? Secondly, the manufacturer should have internal quality control measures in place, and should be able to demonstrate that they work. Your impressions here matter, even if you are a novice at vendor selection. It is important that your vendors and manufacturers are attentive to your business needs and concerns, and conversely, you to theirs. The relationship is one of partnership, and if you get a bad feeling, it probably is not a good fit.

———————◇———————

Lane Musgrave offers some wise reflections on the manufacturing relationship:

"Just like most things in life, success with manufacturing comes down to relationships. For a startup, the priority in finding a CM should not be chasing the cheapest unit costs or best quality parts. These are important considerations but the determinant of success for your product will be a CM who is willing to go to war with you to make sure you both are successful in the venture of manufacturing your widget. What can go wrong, will go wrong. Whether you survive the lost shipments, product recalls, and design flaws will come down to the relationship and trust you have with your CM.

Reserve Strap's road to manufacturing was full of false starts, poor decisions and improvisation. Believe it or not, you will find no shortage of people promising you they can deliver your product quick and cheap because the first people you talk to at the CM are the sales guys who just want you to give them your business. They aren't properly incentivized to get your product to market and the moment you sign the contract, you will be handed off to someone else. The key to judging potential vendors is getting face time with the engineers and project managers who will be responsible for delivering your product.

The biggest challenge for any group will be finding vendors you can trust and the best tool for this is by asking people you trust already for referrals.

I learned this lesson the hard way with some vendors we chose. When you're shopping your product to vendors and contractors, they will tell you that they can do it cheaper than everyone else and faster than everyone else. And that you will get the same service that all their clients get, even though you're a small fish. The only way to judge these claims is to vet references. If a vendor scoffs at a reference request, run for the hills. If you ask for three and they only give you one, run. When you get the references, do the diligence to pursue each one and ask the difficult questions. You will never regret this extra time spent on research before writing a check. It will save you weeks and months of heartache."

Let's hope that next time around vendor selection goes smoother for the mustache comb. We do pretty well on the electrical side. We source our early PCBs from a stateside provider. We pay a little more, but they deliver a quality product on time. And we are able to go through a few revisions quickly. We investigate several paths for the plastic parts. At a projected sales volume for the first year of 3000 units, we need to keep the tooling cost down. The parts may cost a bit more this way, but if we can at least break even that first year we can go for cost reduction in subsequent years. The fastest option we find is a lesser known rapid tooling and production outfit in the US. Again, the part cost is higher, but they promise a low tooling price and a shockingly fast delivery for first shots of fifteen business days. A competing bid from the Chinese factory we found by personal referral has a slightly higher tooling quote, but piece price comes in at a quarter of the price. Great, except that the delivery time for tooling alone is ten weeks. And we are already behind schedule for shipment to customers.

We choose the faster option so that we can get to market quicker. We ask about visiting the factory. They don't seem too keen on that, but we understand they are set up for speed, so we relent. When we ask for references they mention client confidentiality, but they do finally produce two names. We are never able to make contact or get any sort of reply from the references. But they have success stories on the website and, based on the armada of equipment that they have at the factory, we imagine that surely they have the capacity to produce. We sign the contract and send the money, including a fee for expedited delivery to get the parts even faster. They were never responsive by phone, but now, when we ask to clarify the tolerances, there's also a lag of several days between emails. They do get back to us after the tools have been cut, with the explanation that on account of their fast processes, they do not follow the drawing tolerances anyway. We get the parts a week late and they will not refund the fee for expediting. Our original contact disappears and our emails get shuffled through a web of bureaucracy. We need to make some adjustments but communication becomes impossible. The parts are unacceptable to ship and it becomes obvious that this company is incapable of delivering what we need. Now both our schedule and budget are out of control. We jettison this supplier and make the investment to visit both of the factories in China that we have already quoted. We choose the one that came from referrals and end up establishing a solid working relationship, but we have wasted a lot of time and money in the process, and have had to endure countless moments of anxiety and frustration.

It is especially important for a small company or entrepreneur to carefully vet their suppliers. Without the cash reserves that a larger corporation might have, faulty shipments or otherwise failed production can sink a new business before it has even had the chance to get its product to market.

Your insurance against catastrophe starts here in the selection process. Research the management team and the philosophy of potential contract manufacturers. Ask for references and check those references. Visit the factory. Be sure that they want, and need, your business. A committed manufacturer will make a difference in your venture's success.

13

Review and Conclusion: Wrapping up the Mustache Comb

The mustache comb has come a long way in the past fifteen months. If we had it to do all over again, it would go a bit smoother, but we are delivering products now and are hopeful to become profitable next year. We worked with individual consultants instead of a comprehensive development team in order to keep costs down. Under different circumstances the development cost could easily be three times as much. We can tally up what it has cost us to manufacture so far:

Industrial Design:	$10,000
Mechanical Engineering:	$22,000
Electrical Engineering:	$23,000
Prototypes:	$4200
Tooling:	$15,250
Regulatory Testing:	$15,000
Total Development Cost:	**$89,450**

Comb:	$2.25
Cover:	$1.95
PCB: (not yet cost optimized)	$18.00
3000 units:	**$66,600**
Cost per unit:	**$22.20**

Total Capital for Manufacturing: $156,050

In the future, we should be able to get the cost of the PCB down to about a quarter of its current cost which will put us in a more comfortable position with the profit margin of this retail product. Our break-even volume for manufacturing will be at 4002 units when we sell the comb for $39. Take caution that this does not account for shipping expenses, sales and support team, and other business expenses or overhead. Also, as this is an Internet-connected product, there is a related app that can be accessed from a smartphone. Since it does not relate to the actual manufacturing of the product I have left that piece out, but depending on the firm or consultants used, app development could equal or exceed the expense of design and engineering.

Let's review the steps of product manufacturing from our new vantage point. Although we've covered each step within limited portions of this book, each item follows a science of its own that could be the focus of a career. This has been an introduction to the pursuit of specialized manufacturing knowledge that leads to successful products. Greatly simplified, a new product will progress through these stages:

Product Concept and Design
Engineering
Prototypes

Evaluation and Adjustments
Documentation
Request for Quote
Vendor Selection
Tooling Fabrication
Product Fabrication and Assembly
Regulatory Testing
Shipping and Customs
Fulfillment to Customer

Together, these elements comprise the product development branch of a product company. Additional pieces of the puzzle, which include marketing, sales, logistics, distribution, and more, are equally important and demanding. Each of these requires a similar level of devotion to planning and resources as the manufacturing segment that we have been tackling here. Coordinated effort across the disciplines sets a foundation for the ultimate success—a profitable company with products you are proud of.

I've enjoyed creating this manufacturing guide and reflecting on all the little details it takes to create a product. They sure do add up—there are lots of things to juggle and several areas of expertise that must be managed together. Instead of being overwhelming, I hope that this guide prepares you to face your manufacturing challenges and I hope that it provides a jumping off point for great new things. If you are left needing more specific support please write to me at Bright Product Development: bailey@brightpd. com. And if you'd like, you may find updated resources and expanded material at brightpd.com. Now it's time to get to work!

APPENDIX I
Sample Product Schedule

In order to have even the faintest hope of predicting delivery time, it is necessary to construct a detailed schedule. It helps keep track of all the mundane steps in the project, and most importantly, it helps ensure that you do not overlook critical tasks that, if forgotten, would significantly delay completion. This simplified schedule follows the mustache comb through all of the main milestones, starting in January of year one and continuing through to delivery to the customer in March of year two. You can find this schedule in a downloadable format at brightpd.com/schedule.

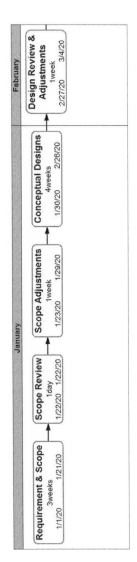

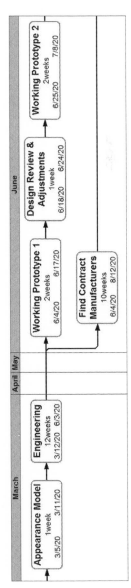

141

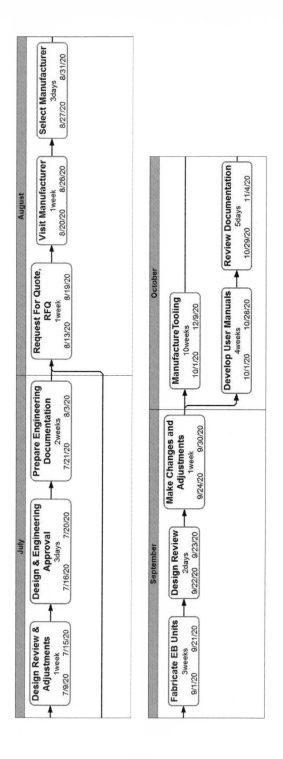

142

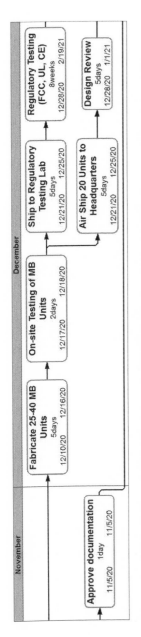

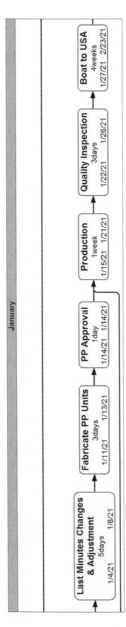

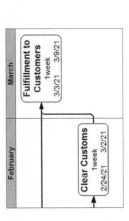

143

APPENDIX II
Plastic Injection Molded Part Design Guidelines

Plastic design principles easily fill entire books. I have attempted to reduce these principles to these most basic guidelines. As guidelines, there is room for them to be stretched or even broken; indeed, I have broken every single one of them in actual production parts. Holding to the guidelines, however, can minimize costly surprises.

1. **Plastic wall thickness should be 1-3mm nominal.** Thinner walls can have filling problems, and thicker walls can have cooling problems that lead to long cycle times and part warpage.

2. **Rib thickness should be 66% or less of the nominal wall thickness. 50% or less is even better.** When a rib is as thick as or thicker than the wall it attaches to, there will likely be a surface imperfection called "sink" on the outside. It can also cause the part to warp.

3. **Avoid undercuts if possible.** Undercuts are common and can be accommodated by an additional action in the tool with a "slide" or a "lifter". However, the tool will be less expensive if you avoid undercuts.

4. **All vertical features should have draft to allow the part to slip out of the mold.**

5. **Draft angle should be 3° or greater if possible.** Cosmetic surfaces are often textured. Textured surfaces require greater draft than glossy surfaces.

6. **Draft on screw bosses, internal ribs, or glossy surfaces may be .5° to 1°.**

7. **Bosses should have support ribs.** Ribs help the material flow during the injection process and also strengthen the boss.

8. **There should be no solid sections of plastic that are thicker than the nominal wall.** Again, this would cause sink and warpage.

9. **No sections should be less than .75 mm in thickness.** Thin sections run the risk of not filling properly and are often manifested as undesirable sharp edges.

APPENDIX III
Standard Drill Bit Sizes

Standard inch drill bits come sized in fractions of an inch, by letter, and by number. Numbered drill bits are for smaller diameters up to .228 inches and letter drill bits are larger—up to .413 inches. Fractional-sized drill bits range from very small to very large, with the most commonly available diameters coming in 1/16 inch increments.

STANDARD INCH DRILL SIZES

DRILL SIZE	DECIMAL INCH	DRILL SIZE	DECIMAL INCH	DRILL SIZE	DECIMAL INCH	DRILL SIZE	DECIMAL INCH
80	.0135	39	.0995	A	.2340	37/64	.5781
79	.0145	38	.1015	15/64	.2344	19/32	.5938
1/64	.0156	37	.1040	B	.2380	39/64	.6094
78	.0160	36	.1065	C	.2420	5/8	.6250
77	.0180	7/64	.1094	D	.2460	41/64	.6406
76	.0200	35	.1100	1/4....E	.2500	21/32	.6562
75	.0210	34	.1110	F	.2570	43/64	.6719
74	.0225	33	.1130	G	.2610	11/16	.6875
73	.0240	32	.1160	17/64	.2656	45/64	.7031
72	.0250	31	.1200	H	.2660	23/32	.7188
71	.0260	1/8	.1250	I	.2720	47/64	.7344
70	.0280	30	.1285	J	.2770	3/4	.7500
69	.0292	29	.1360	K	.2810	49/64	.7656
68	.0310	28	.1405	9/32	.2812	25/32	.7812
1/32	.0312	9/64	.1406	L	.2900	51/64	.7969
67	.0320	27	.1440	M	.2950	13/16	.8125
66	.0330	26	.1470	19/64	.2969	53/64	.8281
65	.0350	25	.1495	N	.3020	27/32	.8438
64	.0360	24	.1520	5/16	.3125	55/64	.8594
63	.0370	23	.1540	O	.3160	7/8	.8750
62	.0380	5/32	.1562	P	.3230	57/64	.8906
61	.0390	22	.1570	21/64	.3281	29/32	.9062
60	.0400	21	.1590	Q	.3320	59/64	.9219
59	.0410	20	.1610	R	.3390	15/16	.9375
58	.0420	19	.1660	11/32	.3438	61/64	.9531
57	.0430	18	.1695	S	.3480	31/32	.9688
56	.0465	11/64	.1719	T	.3580	63/64	.9844
3/64	.0469	17	.1730	23/64	.3594	1	1.0000
55	.0520	16	.1770	U	.3680	1-3/64	1.0469
54	.0550	15	.1800	3/8	.3750	1-7/64	1.1094
53	.0595	14	.1820	V	.3770	1-1/16	1.1250
1/16	.0625	13	.1850	W	.3860	1-11/64	1.1719
52	.0635	3/16	.1875	25/64	.3906	1-7/32	1.2188
51	.0670	12	.1890	X	.3970	1-1/4	1.2500
50	.0700	11	.1910	Y	.4040	1-19/64	1.2969
49	.0730	10	.1935	13/32	.4062	1-11/32	1.3438
48	.0760	9	.1960	Z	.4130	1-3/8	1.3750
5/64	.0781	8	.1990	27/64	.4219	127/64	1.4219
47	.0785	7	.2010	7/16	.4375	1-1/2	1.5000
46	.0810	13/64	.2031	29/64	.4531		
45	.0820	6	.2040	15/32	.4688		
44	.0860	5	.2055	31/64	.4844		
43	.0890	4	.2090	1/2	.5000		
42	.0935	3	.2130	33/64	.5156		
3/32	.0938	7/32	.2188	17/32	.5312		
41	.0960	2	.2210	35/64	.5469		
40	.0980	1	.2280	9/16	.5625		

Metric drill bits come in increments as small as 0.10mm in small diameter bits and increments of up to 0.50 mm in larger diameter bits.

STANDARD METRIC DRILL SIZES

mm	mm	mm	mm	mm	mm
0.10	3.10	6.10	9.10	15.25	27.50
0.20	3.20	6.20	9.20	15.50	28.00
0.30	3.30	6.30	9.30	15.75	28.50
0.40	3.40	6.40	9.40	16.00	29.00
0.50	3.50	6.50	9.50	16.25	29.50
0.60	3.60	6.60	9.60	16.50	30.00
0.70	3.70	6.70	9.70	16.75	30.50
0.80	3.80	6.80	9.80	17.00	31.00
0.90	3.90	6.90	9.90	17.25	31.50
1.00	4.00	7.00	10.00	17.50	32.00
1.10	4.10	7.10	10.20	17.75	32.50
1.20	4.20	7.20	10.50	18.00	33.00
1.30	4.30	7.30	10.80	18.50	33.50
1.40	4.40	7.40	11.00	19.00	34.00
1.50	4.50	7.50	11.20	19.50	34.50
1.60	4.60	7.60	11.50	20.00	35.00
1.70	4.70	7.70	11.80	20.50	35.50
1.80	4.80	7.80	12.00	21.00	
1.90	4.90	7.90	12.20	21.50	
2.00	5.00	8.00	12.50	22.00	
2.10	5.10	8.10	12.80	22.50	
2.20	5.20	8.20	13.00	23.00	
2.30	5.30	8.30	13.20	23.50	
2.40	5.40	8.40	13.50	24.00	
2.50	5.50	8.50	13.80	24.50	
2.60	5.60	8.60	14.00	25.00	
2.70	5.70	8.70	14.25	25.50	
2.80	5.80	8.80	14.50	26.00	
2.90	5.90	8.90	14.75	26.50	
3.00	6.00	9.00	15.00	27.00	

APPENDIX IV
Standard Sheet Metal Thickness

Use standard thicknesses when designing sheet metal parts. Different materials have different standard thicknesses. Check on the availability of specific thicknesses; not all manufacturers will carry all sizes.

STANDARD INCH SHEET METAL THICKNESS

GAGE NUMBER	STEEL INCH	STEEL mm	GAGE NUMBER	STEEL INCH	STEEL MM
3	.2391	6.07	20	.0359	0.91
4	.2242	5.69	21	.0329	0.84
5	.2092	5.31	22	.0299	0.76
6	.1943	4.94	23	.0269	0.68
7	.1793	4.55	24	.0239	0.61
8	.1644	4.18	25	.0209	0.53
9	.1495	3.80	26	.0179	0.45
10	.1345	3.42	27	.0164	0.42
11	.1196	3.04	28	.0149	0.38
12	.1046	2.66	29	.0135	0.34
13	.0897	2.28	30	.0120	0.30
14	.0747	1.90	31	.0105	0.27
15	.0673	1.71	32	.0097	0.25
16	.0598	1.52	33	.0090	0.23
17	.0538	1.37	34	.0082	0.21
18	.0478	1.21	35	.0075	0.19
19	.0418	1.06	36	.0067	0.17

ALUMINUM INCH	MM
.250	6.35
.190	4.83
.160	4.06
.125	3.18
.100	2.54
.090	2.29
.080	2.03
.063	1.60
.050	1.27
.040	1.02
.032	0.81
.025	0.64
.020	0.51
.016	0.41

ACKNOWLEDGEMENTS

I have had invaluable help from many experts in the fields of product development and manufacturing. I'd like to specifically thank my following cohorts, who have each contributed specialized knowledge that makes this book better.

Amber Hagopian, Owner of Brixton Banks
Jack Daniels, President of Eastbridge Engineering
Jessica Campbell, Founder of Pristine Circuits
John Kestner, Principal of SuperMechanical
Kyle Maxey, Contributor to Engineering.com
Lane Musgrave, Co-founder of Reserve Strap
Rich LeGrand, President of Charmed Labs
Song Kosumsuppamala, Chief Engineer of Alen Corporation
Steve Fridley, President of Defender Innovations

BIBLIOGRAPHY

Adams, Robert J. If You Build It Will They Come?, John Wiley and Sons, Inc, 2010.

Anderson, Chris. "Mexico: The New China," The New York Times, 26 January 2013, retrieved 21 December 2016. http://www.nytimes.com/2013/01/27/opinion/sunday/the-tijuana-connection-a-template-for-growth.html

Bralla, James G. Design for Manufacturability Handbook, 2nd Edition, McGraw-Hill, 1999.

Lefteri, Chris. Making It: Manufacturing Techniques for Product Design, Laurence King Publishing, 2007.

Lefteri, Chris. The Plastics Handbook, RotoVision, 2008.

Part and Mold Design, Thermoplastics: A Design Guide, Bayer Material Science, 2000.

"Planet Money Makes a T-Shirt." Retrieved 2 May 2016. http://www.planetmoney.com/shirt

"Polyjet for Injection Molding, Technical Application Guide," Stratasys. Retrieved 18 January 2016. http://bit.ly/2uzSD6X

Malloy, Robert A. Plastic Part Design for Injection Molding, 2nd Edition. Hanser Publishers, March 2010.

"Standard Practice for Coding Plastic Manufactured Articles for Resin Identification." Standard Practice for Coding Plastic Manufactured Articles for Resin Identification. ASTM International. Retrieved 20 October 2016. http://www.astm.org/Standards/D7611.htm

RESOURCES

Find an updated resource list at brightpd.com/resources.

3D Hubs
https://www.3dhubs.com/
3D printing marketplace. Find local 3D printing here, often at a good price.

Bolt
https://bolt.io/
Venture capital for hardware companies coupled with a complete engineering team. Their most immediate and accessible value is in their blog. It is one of the most practical and honest resources I've found for hardware startups.

Blackbox
https://www.blackbox.cool/
A shipping company from the creators of Cards Against Humanity. Outsource your shipping, and your headache, to them.

Core77
http://www.core77.com/
All things design related. Their design directory is one of the best places to search for product design firms by specialty and by region.

Crowd Supply
https://www.crowdsupply.com/
Crowdfunding platform specifically for hardware products. Check out their Providers page for additional product

development resources: https://www.crowdsupply.com/
providers

Dragon Innovation

https://www.dragoninnovation.com/
Factory selection and production oversight. They aid in taking a product from prototype to high volume production.

Eastbridge Engineering

http://www.eb-intl.com/
Quotes and qualifies production facilities in the Asia-Pacific region. They also cover manufacturing launch and supply chain management.

Flex Invention Lab: San Francisco

https://flex.com/connect/innovation-sites/san-francisco-california-invention-lab
Hardware startup lab of the international manufacturer, Flex. You leverage their deep expertise, and they gain insight into emerging technology companies.

Forecast 3D

http://www.forecast3d.com/index.html
Produces 3D printed models using various processes with a specialty in cast urethanes

IDSA

http://www.idsa.org/
The Industrial Designers Society of America is the premier professional society for product designers. Learn about the profession and connect to reputable firms.

Indiegogo
https://www.indiegogo.com
Crowdfunding platform. They have a partnership with Arrow Electronics to provide engineering and production support to select technology product campaigns.

Kickstarter
https://www.kickstarter.com/
Crowd funding platform for projects of all types. Check out their Resources page for more product development service companies: https://www.kickstarter.com/help/resources/

Make: 3D Printers Buyer's Guide
http://makezine.com/comparison/3dprinters/
Online buyer's guide

Maker's Row
https://makersrow.com/
Online directory of American manufacaturers with an emphasis on textiles, fashion, and furniture. They also having learning resources for the clothing and fashion industries.

Make XYZ
https://www.makexyz.com/
3D printing marketplace. Find local 3D printing here, often at a good price.

PCH, Highway 1
http://highway1.io/
A hardware startup accelerator that is a division of the international manufacturer, PCH.

Penn Engineering

http://www.pemnet.com/

Provides threaded inserts and other specialty inserts for sheet metal and plastic applications

Proto Labs

http://www.protolabs.com/

Quick-turn injection molding, machining, and 3D printing

Rapid Sheet Metal

http://www.rapidmanufacturing.com/

Quick-turn sheet metal and machined prototypes. Now acquired by Proto Labs.

RP Prototype

http://rpgroupltd.com/

Prototypes and low-volume production at a good price. China based and Western operated.

Shapeways

http://www.shapeways.com/

3D printing of all sorts. They also maintain an on-line marketplace of parts.

Sierra Circuits

https://www.protoexpress.com/

PCB source for prototype quantities and larger quantities. They have an on-line quoting system.

Stratasys Direct Manufacturing

https://www.stratasysdirect.com/
Combines the expertise of three 3D printing companies: Solid Concepts, Harvest Technologies, and RedEye. My choice for SLS parts.

Tenere

http://www.tenere.com/rapid-prototyping/
Quick-turn 3D printing and high quality SLA

NOTES

All photographs are by Bailey Briscoe Jones unless otherwise noted.

[1] Photo by Britiju. Public domain.

[2] Photo by Chris Harrison. Photo has been cropped. Used under Creative Commons license CC BY-SA, 2.0 http://creativecommons.org/licenses/by-sa/2.0/legalcode

[3] Photo by Mike1024. Public domain.

[4] Image from http://www.packaginggraphics.net/plastic-recycle-logos.htm, retrieved 18 October 2016.

[5] Case Study, in a different format, originally appeared in: Models for Success: When, How and Why Designers are Using 3D Printing Today, Made for Me, August 2016.

GLOSSARY

3D Printing Any computer controlled additive process used to build 3 dimensional objects

Additive Manufacturing 3D printing performed on an industrial scale

Additive Process A fabrication method that adds elements together to achieve the final result. 3D printing is an additive process.

Bill of Materials List of mechanical components or electrical components that comprise a product

Blow Molding A manufacturing process where molten plastic is expanded like a balloon into a mold to produce bottles and other hollow shapes

Bridge Tooling Limited production tooling that is meant to be replaced as manufacturing volumes increase. Soft tooling often serves as bridge tooling.

CAD Computer Aided Design usually refers to drawings and three dimensional computer models that are used to document and manufacture parts.

CAM Computer Aided Manufacturing. CAD files can be used to control automated manufacturing equipment.

Casting Includes sand casting, die casting, and investment casting. A manufacturing process where molten metal is poured into a mold.

Cast Urethane See RTV.

CM Contract Manufacturer

CNC Computer Numerical Control. Automated machining or other manufacturing process that is controlled by a computer with minimal human intervention.

Compression Molding A manufacturing process often used with thermoset materials such as rubber. A pre-measured putty-like plug or preform is inserted into a mold and then cured under heat and pressure.

DFM Design for Manufacturing

DMLS Direct Metal Laser Sintering. A 3D printing process that fuses together powdered metal with a laser

Draft The small angle on vertical faces of injection molding tools that allows the part to slip out of the mold.

Drill File A file that controls the location and size of drilled holes in a PCB

Durometer A measure of the hardness of rubber. Consumer products are normally specified on the Shore A scale with 10 being extremely soft and 95 being quite hard.

Extruding A manufacturing process where molten plastic or metal is forced through a die to produce a continuous shape such as an "L" or a tube. The continuous part is then cut to length.

FDM Fused Deposition Modeling. A 3D printing process that extrudes a filament of material (often ABS or PLA plastic) through a hot nozzle.

First Article Inspection This is a formalized process of inspecting the first parts off a manufacturing line. Parts are evaluated according to drawings, golden samples, or other specifications.

First Shots These are the first injection molded parts produced by the factory. First shots are used to evaluate and improve injection molding parameters.

Forging A manufacturing method that shapes metal with repeated hammer blows until the material takes the shape of the die

FR4 Flame retardant fiberglass composite. This is the most common substrate for printed circuit boards.

Gerber 2D files that are used to describe PCB layer images such as copper layers and solder masks

Golden Sample This is an actual part or assembly that production parts can be measured against. A golden sample is often used as the color standard for painted parts or injection molded parts.

Hard Tooling Injection molding tools made from hardened tool steel. It is more costly than soft tooling, but supports manufacturing volumes in the hundreds-of-thousands range at a lower piece price than soft tooling.

Injection Molding A manufacturing process where molten plastic is injected into a mold to produce parts of various shapes. Injection molding is the most common plastic manufacturing method.

Lifter A lifter provides a means for accommodating undercuts in injection molded parts. A portion of the tool is hinged to the main tool body and automatically swings free of the undercut with the motion of opening the tool.

Living Hinge A hinge that is integral to the actual material. Plastic parts may be molded with a very thin section that acts as a hinge.

Manufacturing Build These products are usually built in a quantity of 100 or less to test and optimize the manufacturing and assembly process. The test units are then evaluated according to the specifications and quality requirements before proceeding to pilot production.

MBD Model Based Definition captures tolerances and required manufacturing notes (that traditionally have been conveyed with 2D drawings) within the 3D CAD file itself.

MO Minimum Order dollar amount that is available to order from a contract manufacturer

MOQ Minimum Order Quantity is the minimum batch size that is available to order from a contract manufacturer.

Netlist A file that describes the terminals and their connections for an electrical design

NRE Non-Recurring Engineering is the cost related to setting up production. Tooling cost, for example, is sometimes grouped together with the NRE cost.

Parison A tube of molten plastic that is prepared to be blow molded by expanding it with forced air into a mold

PCB Printed Circuit Board

Pick and Place An automated machine that places surface mount components on a PCB

Pilot Production The first real products that are produced by the factory. Production is usually limited to hundreds and the process is further optimized before proceeding to full-scale mass production.

Polyjet A 3D printing process that deposits a liquid material in layers using print-heads, much like those in a bubble jet printer for paper. The liquid layers are cured with UV light.

Rapid Manufacturing An inclusive term that encompasses CNC machining, additive manufacturing, and other subtractive or additive processes, normally automated and controlled by a computer

Rapid prototyping An automated process for producing prototypes that includes 3D printing, CNC machining, and other computer-controlled fabrication methods.

RFQ Request for Quote. This package includes product specifications, engineering documentation, quantity forecasts, and the company's sales and marketing position.

RTV Room temperature Vulcanization, also called Cast Urethane. A prototyping method where liquid urethane is poured and cured in a flexible silicon mold

Sink A deformation in injection molded parts. Sink occurs when uneven cooling causes the material to "sink" in and appear as a depression in the surface.

SLA Stereo Lithography (Apparatus). A 3D printing process that cures a liquid resin with UV light, usually by way of a laser beam

Slide A slide provides a means for accommodating undercuts in injection molded parts. The slide is pulled out from the undercut separately from the main tool motion before the tool is opened.

SLS (SL) Selective Laser Sintering. A 3D printing process that fuses together a powdered material (often nylon plastic) with a laser

SMT Surface Mount Technology is a system of placing an electrical component on a PCB solder pad as an automated alternative to using through holes.

Soft Tooling Injection molding tools made out of easy-to-machine metal such as aluminum or sometimes carbon steel. Soft tooling can be made faster and cheaper than hard tooling and can support manufacturing volumes in the tens-of-thousands range.

STL Stereolithography (File). This is the most common file format used for producing 3D printed parts.

Subtractive Process A fabrication method that removes material to achieve the final result. Machining is a subtractive process.

Thermoplastic A plastic type that softens and becomes moldable with heat and solidifies when cooled. Thermoplastics may be re-melted.

Thermoforming A manufacturing process where a heated plastic sheet is draped over a pattern and pressed into shape

Thermoset Plastic A plastic or rubber that cures with heat or chemical reaction. Thermoset plastics "set" and cannot be re-melted.

Tooling Specialized pieces of hardware such as injection molds or extrusion dies that are made for the manufacture of custom parts. Tooling allows for efficient production of parts in high volumes.

Tool-safe Also called steel-safe. Changes to a tool may require removing material or adding material. A change that requires removing tool material is much easier (the material can simply be machined away) and is called a tool-safe change.

TPE Thermoplastic Elastomer. This rubber-like, injection-moldable thermoplastic is an easier-to-process alternative to traditional thermoset rubber.

UNC United National Coarse is the standard designation in the United States for coarse screw threads. See also UNF.

Undercut A plastic overhanging feature that would prevent an injection molded part from being slipped straight out of the mold. Undercuts require side action in the tool.

UNF United National Fine is the standard designation in the United States for fine screw threads. See also UNC.

INDEX

ABOUT THE AUTHOR

Photograph by Gino Verna

Bailey Briscoe Jones is the founder of Bright Product Development, a product design firm located in Austin, Texas. He is a consumer product development veteran with experience in the United States and Scandinavia and has worked with industry leaders such as Dell, Oakley, Vestas, and 3D Systems. He helped pioneer the integration of electronics into sunglasses (the Oakley Thump mp3 player) before we knew the term "wearable electronics" and was designing 3D printers for 3D Systems back when it was still called "rapid prototyping."

Bright Product Development collaborates with technology companies to engineer profitable products in fields that range from wearable electronics to home appliances. Find out more about Bailey and Bright Product Development at brightpd.com.

57692020R00105

Made in the USA
Columbia, SC
11 May 2019